Peacock Publishing Presents

AF884694

Cryptocurrency: Decrypted

A Practical Guide to Crypto Ownership, Wallets, and Real Security

By Sean Carey

Cryptocurrency, Decrypted
A Practical Guide to Ownership, Wallets, and Real Security

Second Edition

First edition published 2020
Second edition published 2026

ISBN: 978-1-944963-03-3 (paperback)
ISBN: 978-1-944963-04-0 (ebook)

Published by Peacock Publishing
A subsidiary of WillowHeart, LLC

Printed in the United States of America

INTRODUCTION

This book was first published in 2020, during one of the most explosive periods in the history of cryptocurrency and decentralized finance. At the height of the DeFi boom, hundreds of new tokens and coins were launching every single day. Innovation was moving at breakneck speed, barriers to entry were lower than ever, and for many people, crypto represented their first real encounter with true financial sovereignty.

With that opportunity came an uncomfortable truth: most participants entered the space without a foundational understanding of how to protect what they were building.

This is not a how-to book. There are countless guides that explain how to buy a token, connect a wallet, or chase the next promising project. This is a why-to book. While anyone can get started in crypto in a matter of minutes, protecting digital assets—especially after early gains—requires a very different mindset.

Crypto does not behave like traditional finance. There is no customer service line to call, no fraud department to reverse mistakes, and no institution standing between you and a bad decision. In decentralized systems, ownership is absolute—and so is responsibility. The same tools that grant freedom and autonomy can just as easily expose users to irreversible loss if they are not properly understood.

This book explores the often-overlooked layers of personal asset protection in the crypto ecosystem. Rather than focusing on price charts or market speculation, it examines the foundations that keep digital assets secure: your home network, your computing environment, your operational habits, and ultimately, your relationship with self-custody.

The goal is not fear, but awareness. Not paranoia, but preparedness. True decentralization demands a shift in thinking—from reliance on institutions to personal accountability and informed self-custody. Whether you are new to crypto or have already experienced meaningful gains, understanding why security matters is the first step toward protecting what you've earned.

In the chapters that follow, we move beyond hype and into fundamentals—because in crypto, freedom is only as strong as the systems that safeguard it.

A NECESSARY NOTE ON COMPUTER SECURITY

This book is not intended to be a comprehensive guide to computer or network security. Entire careers—and countless excellent books—are dedicated to those subjects alone, often exploring them at a level of technical depth far beyond the scope or intent of this work. Operating system hardening, enterprise-grade intrusion detection, advanced malware analysis, and formal security architecture are complex disciplines that deserve focused, specialized treatment.

However, it is impossible to have an honest conversation about cryptocurrency security without addressing key aspects of computer security.

Cryptocurrency does not exist in isolation. Every wallet, transaction, private key, and authentication step ultimately passes through a computing device and a network. If the underlying system is compromised—through malware, phishing, keylogging, browser exploits, or poor network hygiene—then even the most carefully chosen wallet or blockchain protocol becomes irrelevant. A perfectly secure blockchain does not protect a user whose computer is already under an attacker's control.

For this reason, the security concepts discussed in this book intentionally overlap with foundational computer security principles. The goal is not to turn the reader into a cybersecurity professional, but to establish enough practical awareness to prevent the most common and devastating failures. Many crypto losses occur not because of flaws in cryptographic systems, but because attackers exploit weak passwords, unpatched software, insecure email accounts, compromised browsers, or poorly secured home networks. These are not theoretical risks—they are the dominant attack vectors in real-world crypto theft.

Cryptocurrency shifts responsibility away from institutions and places it squarely on the individual. When you become your own bank, you also inherit responsibilities that banks traditionally manage behind the scenes: device security, access control, transaction verification, and threat awareness. Ignoring the security of the systems used to access crypto is equivalent to installing a vault door while leaving the building unlocked.

Accordingly, this book addresses computer security only to the extent necessary to support safe cryptocurrency usage. The focus remains practical and purpose-driven: protecting private keys, preventing unauthorized access, reducing exposure to common attack techniques, and building layered defenses that make exploitation difficult and unattractive to attackers. Readers seeking deeper technical mastery are strongly encouraged to continue their education through dedicated cybersecurity resources—but the principles presented here form a critical baseline.

In short, while this is not a computer security manual, cryptocurrency security cannot exist without it. Understanding—and respecting—that relationship is essential for anyone who intends to hold, manage, or grow digital assets responsibly.

Readers who already possess deep familiarity or professional experience in computer security, networking, or related disciplines may find that some of the material in these sections revisits concepts they already understand. In those cases, portions can certainly be skimmed or skipped. That said, cryptocurrency security is an interconnected system, and even experienced users may encounter subtle overlaps, assumptions, or implementation details that differ from traditional security models. Skipping foundational sections entirely risks missing how these concepts specifically apply to crypto—where responsibility, irreversibility, and self-custody change the stakes. For that reason, even seasoned practitioners may benefit from viewing these topics through the lens of digital asset security rather than conventional computing alone.

CHAPTER 1 - INTRODUCTION TO CRYPTO AND CRYPTOCURRENCY

WHAT IS CRYPTOCURRENCY?

Cryptocurrency is a form of digital currency that operates without reliance on central banks or trusted third parties to verify transactions or issue new units of currency. Instead, it uses cryptographic techniques to secure transactions and record them on a publicly distributed ledger known as a blockchain. While this concept may seem complex at first, it becomes clearer as the underlying ideas are explored.

Today, there are thousands of cryptocurrencies in circulation, each designed with its own purpose and use case. The first and most well-known cryptocurrency, Bitcoin, was introduced in 2009 by an anonymous individual or group using the pseudonym Satoshi Nakamoto.

In 2008, Nakamoto published a white paper titled Bitcoin: A Peer-to-Peer Electronic Cash System, outlining the foundational principles of blockchain technology. This innovation made it possible for digital currency to function similarly to government-issued (fiat) money—without requiring intermediaries such as banks, payment processors, or financial institutions to verify transactions.

One of the most important problems blockchain technology solves is known as the double-spending problem, a fundamental challenge associated with digital cash. Because digital data can be copied easily, a secure system is required to ensure that currency units cannot be duplicated or spent more than once. Blockchain addresses this issue by maintaining a tamper-resistant, chronological record of transactions that is shared across a network of participants.

Unlike traditional financial systems, where central authorities establish and enforce transaction legitimacy, cryptocurrencies rely on decentralized networks to validate and record all activity. Once transactions are confirmed and added to the blockchain, they become effectively immutable—meaning they cannot be altered without overwhelming consensus. This immutability enhances security, reduces fraud, and allows trust to emerge from the system itself rather than from any single institution.

To understand this more intuitively, imagine a small group of friends who agree to keep a shared financial record. Instead of trusting one person to maintain the ledger, each friend keeps an identical copy. When a transaction occurs, everyone records it at the same time. If someone later attempts to change their copy, the discrepancy is immediately obvious because it no longer matches the others. Trust is not placed in any one individual, but in the shared agreement of the group.

This is, in essence, how a blockchain works. Each participant maintains a copy of the ledger, and accuracy is preserved through transparency, verification, and consensus rather than authority. The system does not rely on trust in people, but on mathematics, shared records, and the inability to quietly rewrite history. While vastly more complex in practice, this principle underlies the blockchain technology used by Bitcoin and many other cryptocurrencies today.

HOW MANY CRYPTOCURRENCIES EXIST?

The number of cryptocurrencies has grown exponentially over the years. As of late 2020, CoinMarketCap reported approximately 13,669 distinct cryptocurrencies. By 2026, that figure had ballooned to an estimated 36.4 million coins and tokens. It is important to note, however, that this number is heavily inflated. As new projects are created, many quickly become inactive, abandoned, or obsolete, causing the total count to fluctuate constantly. Even accounting for this high level of turnover, it is fair to say that tens of millions of cryptocurrencies have existed—highlighting both the explosive growth and the overwhelming noise within the crypto ecosystem.

One of the primary reasons for this rapid expansion is the ease with which new cryptocurrencies can be created. Platforms such as Ethereum provide the infrastructure for developers to write and deploy smart contracts directly on the blockchain. Rather than building an entirely new blockchain from the ground up, developers can leverage existing networks to launch their own tokens with relatively minimal effort. While this accessibility has fueled innovation, it has also contributed to a flood of low-effort, short-lived projects, making discernment and due diligence increasingly important for participants in the space.

HOW CRYPTOCURRENCY WORKS

To make a cryptocurrency transaction, you use a cryptocurrency wallet. Despite the name, a wallet does not physically store digital currency. Instead, it functions as an interface that allows you to view and interact with funds recorded on the blockchain—the public ledger where all transactions are tracked. Your wallet "points" to a specific blockchain address that reflects your balance and transaction history.

Each wallet is associated with a pair of cryptographic keys: a public key and a private key. The public key (or derived wallet address) can be shared with others to receive funds, while the private key authorizes transactions and proves ownership of the assets associated with that address. Together, these keys allow you to send, receive, and securely manage cryptocurrency. Control of the private key is what ultimately determines control of the funds.

BUYING, SELLING, AND USING CRYPTOCURRENCY

Most cryptocurrency transactions begin on cryptocurrency exchanges, which are online platforms that allow users to buy, sell, and trade digital assets. These exchanges typically hold deposits in both fiat currencies (such as USD or EUR) and cryptocurrencies, crediting or debiting user balances as trades are executed. This system allows transactions to occur quickly and efficiently, especially for new users entering the market.

Beyond trading, cryptocurrency can also be used to purchase goods and services. In these cases, a transaction involves authorizing the transfer of a specific amount of cryptocurrency from your wallet address to the recipient's wallet address. Once confirmed on the blockchain, the transaction is recorded permanently, reflecting the movement of funds without the need for a bank, payment processor, or other intermediary.

TRANSACTION PROCESSING AND BLOCKCHAIN VALIDATION

Each cryptocurrency transaction is authorized using the sender's private key and then broadcast to the blockchain network. Network participants—commonly referred to as miners in Proof-of-Work systems or validators in Proof-of-Stake systems—use the corresponding public key to confirm that the transaction was legitimately signed by the holder of the private key. (Proof of Work and Proof of Stake will be covered later in this chapter.)

Once verified, the transaction is grouped with others and added to a block on the blockchain. This process ensures that the public ledger remains accurate, transparent, and effectively immutable, reflecting updated balances for both the sender and the recipient.

Because blockchain transactions are processed by decentralized software rather than centralized institutions, they do not rely on banks or third-party intermediaries. This is what allows cryptocurrency to function as a truly peer-to-peer financial system.

To visualize this process, return again to the example of a group of friends who each keep an identical copy of a shared ledger. Imagine that every friend has a unique private seal, and that each seal leaves a mark that can only be validated under a special blacklight. When someone wants to record a transaction, they stamp the ledger with their private seal. No one else can recreate that mark, but everyone in the group owns the same blacklight. When a new entry appears, each friend independently shines the blacklight on the seal. If the correct hidden pattern appears, they know the transaction could only have come from the rightful owner—even though they never saw or handled the private seal itself.

This is how private and public keys function in a blockchain system. The private key creates the cryptographic signature, while the public key acts like the blacklight—allowing the network to verify authenticity without exposing the secret used to create it. Validation does not require trust in a person or central authority, only confirmation that the cryptographic proof is genuine. In this analogy, the group of friends performing the verification represents the miners or validators who collectively secure the network.

WHY IS IT CALLED A BLOCKCHAIN?

A blockchain is named after its underlying structure—a chain of blocks that store transaction data in a decentralized and secure manner. Each block contains a collection of verified transactions, such as records showing that one party sent cryptocurrency to another or that funds were received by a specific address. Rather than existing in isolation, these blocks are linked together in a precise order, forming a continuous and transparent history of transactions.

Every block also includes a cryptographic reference to the block that came before it. This reference acts like a digital fingerprint, permanently tying each block to its predecessor and creating an unbroken sequence of records. As new blocks are added, the chain grows longer, with each addition reinforcing the integrity of the entire ledger.

HOW BLOCKCHAIN PREVENTS TAMPERING

This chained structure is what gives blockchain its powerful resistance to tampering. Because each block depends on the one before it, altering a single transaction would require rewriting not only that block, but every block that follows. An attacker attempting to modify past data would need to recalculate and replace the entire chain across the network simultaneously—a task that is computationally impractical under normal circumstances.

This design makes blockchain records effectively immutable. Once transactions are confirmed and added to the chain, they cannot be quietly changed or erased. Instead of relying on trust in a central authority, blockchain relies on structure, cryptography, and distributed consensus to preserve accuracy and security.

HOW BLOCKS ARE ADDED TO THE BLOCKCHAIN

New blocks are added to a blockchain through a consensus mechanism—a process that allows participants in a decentralized network to agree on which transactions are valid and in what order they should be recorded. Consensus ensures that no single entity controls the ledger and that all copies of the blockchain remain synchronized. While many consensus models exist, two of the most widely used are Proof of Work and Proof of Stake.

1. Proof of Work (PoW) – Mining-Based Consensus

Proof of Work relies on miners who use specialized computers to solve complex mathematical puzzles. These puzzles serve as a competitive process for validating transactions and securing the network. When a miner successfully solves the puzzle, they earn the right to add the next block to the blockchain.

In exchange for this work, miners are rewarded with newly issued cryptocurrency and transaction fees. Because solving these puzzles requires significant computational power, Proof of Work makes the blockchain highly resistant to attacks. To alter a previously confirmed block, an attacker would need to control more than 50 percent of the network's total computing power—an undertaking that is economically and practically prohibitive for large, established blockchains.

2. Proof of Stake (PoS) – Stake-Based Consensus

Proof of Stake takes a different approach by removing the need for energy-intensive computation. Instead of miners competing to solve puzzles, validators—sometimes called stakers—are selected to create new blocks based on the amount of cryptocurrency they have committed, or "staked," as collateral.

Validators with larger and longer-held stakes generally have a higher probability of being chosen to validate transactions. This method dramatically reduces energy consumption compared to Proof of Work while still maintaining network security. Modern Proof-of-Stake systems also include safeguards designed to prevent wealthy participants from gaining disproportionate influence or manipulating transaction validation, helping preserve fairness and decentralization.

WHY A BLOCKCHAIN IS SECURE AND RELIABLE

Blockchain technology is considered secure and reliable because its underlying design combines cryptography, decentralization, and consensus. Both Proof of Work and Proof of Stake play critical roles in maintaining the integrity of the network by ensuring that transactions are validated honestly and recorded consistently across all participants.

Once a transaction is confirmed and added to the blockchain, it becomes effectively immutable. Altering past records would require rewriting large portions of the chain and gaining control over a majority of the network's validating power—an undertaking that is economically and computationally impractical for established blockchains.

Decentralization further strengthens security by removing single points of failure. Instead of relying on one central authority to approve or store transactions, blockchain networks distribute this responsibility across thousands of independent participants. This makes censorship, manipulation, or coordinated fraud far more difficult to achieve.

Taken together, these characteristics make blockchain a fundamentally secure system for recording and verifying transactions. Rather than relying on trust in institutions, blockchain relies on transparent rules enforced by code and collective consensus. This trustless design is what makes blockchain one of the most transformative innovations in digital finance.

WHY IS CRYPTO SO POPULAR?

Cryptocurrency has gained widespread popularity due to a combination of growing real-world utility, attractive investment potential, and a broader vision for the future of finance. As the industry continues to evolve, new applications, increasing business adoption, and ongoing technological innovation have pushed crypto from a niche experiment into mainstream financial discussion.

1. Increasing Utility

One of the primary drivers behind cryptocurrency's rise is its expanding range of practical applications. What began as a peer-to-peer digital payment concept has grown into an entire ecosystem built on blockchain technology. Innovations such as decentralized finance (DeFi), blockchain-based gaming, and non-fungible tokens (NFTs)* have introduced new ways to use and interact with digital assets beyond simple buying and selling.

At the same time, real-world adoption continues to increase. More retailers and service providers are beginning to accept cryptocurrency as a form of payment, allowing users to transact without relying exclusively on traditional banking infrastructure. This growing utility has helped reposition cryptocurrency as not just a speculative asset, but a functional tool within a rapidly evolving digital economy.

***A note on NFT's**

Non-fungible tokens, commonly referred to as NFTs, have developed a poor reputation in the public imagination, largely due to waves of mass-produced digital artwork being sold for exorbitant prices. While these speculative uses dominated headlines, they obscured the far more important innovation underlying NFTs.

At their core, NFTs are simply unique, verifiable tokens recorded on a blockchain. This makes them well suited to represent ownership of one-of-a-kind assets. A vehicle title, a property deed, or a certificate of authenticity for a rare collectible could all exist as NFTs. In this model, transferring ownership no longer requires a centralized third party such as a title company, registry, or broker. Instead, the NFT itself represents the asset, and transferring it on the blockchain constitutes a provable change in ownership.

This is the real power of NFTs—not speculative images, but trustless and verifiable ownership of unique items, secured by cryptography and recorded on a public ledger that cannot be quietly altered or disputed.

2. Attractive Investment Opportunity

Cryptocurrency has also emerged as a high-growth asset class, evolving rapidly from a fringe technological experiment into a globally recognized financial sector. Over the past several years, its total market value has surged, drawing the attention of institutional investors, governments, and mainstream media alike.

For many participants, the appeal lies in crypto's potential for outsized returns compared to traditional assets, particularly during periods of rapid innovation or adoption. At the same time, this growth has been accompanied by significant volatility. Dramatic price swings can amplify both gains and losses, making cryptocurrency simultaneously enticing and risky. This combination of opportunity and uncertainty is precisely what makes crypto so compelling—and so misunderstood—underscoring the importance of approaching it not merely as an investment, but as a new financial paradigm that demands informed participation and responsible asset protection.

3. Vision for the Future

Beyond utility and investment potential, many advocates view cryptocurrency as a foundational shift in how money and financial systems operate. Businesses across a wide range of industries are actively exploring blockchain technology as a way to improve efficiency, increase transparency, and reduce reliance on centralized intermediaries.

While the industry remains in its early stages, continued development and adoption suggest a long-term transition toward digital finance. Whether this transformation unfolds gradually or through rapid disruption, cryptocurrency represents more than a passing trend—it reflects a growing desire for financial systems that are more open, programmable, and globally accessible.

BIGGEST CRYPTOCURRENCIES

The cryptocurrency market is highly dynamic, with values and rankings that can change rapidly based on price movements, adoption, and broader market conditions. While individual projects may rise or fall over time, a small group of cryptocurrencies has consistently remained at the top by market capitalization.

As of January 2026, the five largest cryptocurrencies by market capitalization were:

Rank	Cryptocurrency	Market Capitalization*
1	Bitcoin (BTC)	$1.87 trillion
2	Ethereum (ETH)	$401 billion
3	Tether (USDT)	$187 billion
4	XRP	$130 billion
5	BNB	$129 billion

**Market capitalizations are subject to change based on market fluctuations and are provided for illustrative purposes.*

Among these, Bitcoin remains the most valuable and widely recognized cryptocurrency. Its dominance is largely attributed to its first-mover advantage, broad global adoption, and strong network security. Bitcoin is often viewed as the benchmark against which the rest of the crypto market is measured.

Ethereum holds the second position and serves as the foundational platform for decentralized applications, smart contracts, and a vast ecosystem of tokens. Many blockchain projects rely on Ethereum's infrastructure, contributing to its continued relevance and resilience despite market volatility.

While rankings beyond the top positions can change frequently, Bitcoin and Ethereum have consistently maintained their dominance due to their established use cases, deep liquidity, and widespread adoption. Together, they represent a significant portion of the total cryptocurrency market and play a central role in shaping the broader ecosystem.

ADVANTAGES AND DISADVANTAGES OF CRYPTOCURRENCY

Like any major financial innovation, cryptocurrency offers meaningful advantages alongside important risks. Understanding both sides is essential for anyone considering participation in the crypto ecosystem, particularly because responsibility and accountability rest largely with the user.

Advantages

One of cryptocurrency's most notable advantages is speed. Transactions can often be confirmed within minutes, even across international borders. In contrast, traditional financial transfers—especially cross-border payments—may take several days to settle due to intermediary banks and legacy systems.

Cryptocurrency transactions can also be less expensive than traditional banking services. While fees vary depending on the network and level of congestion, sending funds internationally is often significantly cheaper than using wire transfers or remittance services, which frequently involve multiple fees and unfavorable exchange rates.

Another key benefit is financial inclusion. Unlike traditional banking systems, cryptocurrency does not require credit checks, formal identification, or approval from a financial institution. Anyone with an internet connection can create a wallet and participate, making crypto accessible to individuals in regions with limited or unreliable banking infrastructure.

Cryptocurrency also offers enhanced security and privacy when used correctly. Transactions are secured through cryptographic encryption, and users are not required to share personally identifiable information to transact. While blockchain activity is transparent, identities are typically pseudonymous, offering a different privacy model than traditional finance.

Disadvantages

Despite these advantages, cryptocurrency carries significant risks. One of the most important is the lack of insurance protection. Unlike bank deposits, which may be insured by government-backed programs, cryptocurrency holdings are generally not protected. If funds are stolen, lost, or mismanaged, there is often no mechanism for recovery.

Another critical drawback is transaction irreversibility. Once a transaction is confirmed on the blockchain, it cannot be canceled or reversed. Sending funds to the wrong address or falling victim to a scam can result in permanent loss, with no customer support or dispute resolution process.

Cryptocurrency also places full responsibility for access on the user. Wallets are controlled by private keys, and losing a private key means losing access to the associated funds permanently. There is no password reset, recovery hotline, or institution that can restore access.

Finally, cryptocurrency markets are highly volatile. Prices can fluctuate dramatically in short periods, with swings of ten percent or more in a single day not uncommon. While this volatility contributes to crypto's appeal as a high-growth asset, it also makes it a high-risk investment unsuitable for those unprepared for rapid market changes.

HOW TO MINE CRYPTOCURRENCY

Mining is the process by which transactions are validated and added to certain blockchains, most notably those that use a Proof-of-Work consensus mechanism. Miners contribute computing power to verify transactions, secure the network, and maintain the integrity of the public ledger. In return for this work, they are rewarded with newly issued cryptocurrency and transaction fees.

At a basic level, mining requires specialized hardware capable of performing large numbers of calculations efficiently. Two primary types of hardware are commonly used. Graphics Processing Units (GPUs), often found in gaming computers, are flexible and can mine a variety of cryptocurrencies. Application-Specific Integrated Circuits (ASICs), by contrast, are custom-built machines designed to mine a single cryptocurrency or algorithm. ASICs are far more efficient but lack flexibility and are typically more expensive.

Before mining can begin, a cryptocurrency wallet must be set up to receive mining rewards. This wallet functions as the destination address where newly earned coins are deposited once blocks are successfully validated.

Miners must also install specialized mining software that connects their hardware to the blockchain network. This software coordinates the mining process, communicates with other network participants, and submits completed work for verification.

One of the most significant considerations in mining is energy consumption. Mining operations require substantial electricity and generate significant heat, making proper cooling and power efficiency critical. Electricity costs are often the determining factor in whether mining is profitable, particularly as network difficulty increases over time.

Finally, mining profitability is closely tied to market conditions. Because rewards are paid in cryptocurrency, fluctuations in market prices can dramatically affect whether mining operations remain financially viable. For this reason, miners must continuously monitor both operational costs and market trends to assess long-term sustainability.

ARE CRYPTOCURRENCIES A GOOD INVESTMENT?

Cryptocurrencies are neither inherently good nor bad investments. Their suitability depends largely on an individual's risk tolerance, investment goals, time horizon, and understanding of the market. Like any emerging asset class, crypto presents both unique opportunities and meaningful risks, making thoughtful evaluation essential before committing capital.

One of the most important considerations is volatility. Cryptocurrencies are among the most volatile assets available, with prices capable of experiencing extreme swings over short periods of time. While this volatility can create opportunities for significant gains, it also increases the likelihood of rapid and substantial losses, particularly for unprepared investors.

Portfolio diversification is another key factor. Some investors view cryptocurrency as a high-risk, high-reward asset that may complement more traditional investments. When used carefully, allocating a relatively small portion of a diversified portfolio to crypto may help balance overall risk while providing exposure to potential upside. However, overexposure can magnify losses during market downturns.

It is also important to recognize that cryptocurrencies lack the long-term historical data available for traditional assets such as stocks or bonds. Without decades of performance history across multiple economic cycles, predicting long-term outcomes is inherently more uncertain. This makes education and risk management especially important for those entering the space.

Despite these challenges, many institutional investors, financial firms, and industry leaders believe that cryptocurrency may play a role in modern investment portfolios. That said, participation should be grounded in a clear understanding of the technology, the risks involved, and the responsibilities that come with self-custody and decentralized systems.

FINAL THOUGHTS

Cryptocurrency's popularity continues to grow due to its expanding utility, investment potential, and its role in shaping the future of finance. While it offers compelling advantages, it also carries risks that cannot be ignored. Investors who choose to participate should do so with education, caution, and deliberate planning—recognizing that success in this space depends as much on preparation and security as it does on market performance.

CHAPTER 2 - WHY YOU NEED TO SECURE YOUR CRYPTOCURRENCY

WHY YOU NEED TO SECURE YOUR CRYPTOCURRENCY HOLDINGS

If you are wondering why securing your cryptocurrency holdings is so important, consider this: the world of crypto is often compared to the old Wild West—and for good reason. Unlike traditional finance, where banks and institutions provide multiple layers of protection, the cryptocurrency space places far more responsibility directly on the individual. In many cases, there is no safety net beyond the precautions you take yourself.

Much like the Wild West era, where stagecoach robberies and bank hold-ups were common, the only way to truly protect your money or possessions was through personal vigilance and preparation. There were no centralized systems capable of reversing theft or recovering losses after the fact. In many ways, cryptocurrency operates under similar conditions: ownership is direct, control is absolute, and mistakes or breaches can be permanent.

While regulatory agencies continue to explore oversight and consumer protections, cryptocurrencies were originally designed to be decentralized—operating outside traditional banking systems and minimizing reliance on centralized authorities. This decentralization creates powerful opportunities for financial autonomy and freedom, but it also introduces real risks. Understanding those risks, and taking responsibility for securing your assets, is not optional in crypto—it is fundamental.

CENTRALIZED VS. DECENTRALIZED FINANCE: UNDERSTANDING THE DIFFERENCE

Not all cryptocurrency activity is truly decentralized. Many large exchanges—such as **Coinbase**, **KuCoin**, **Kraken**, and **Binance**—operate as *Centralized Finance (CeFi)* platforms. These services act as custodians of user funds, holding cryptocurrencies on behalf of their customers. This custodial model lowers the barrier to entry and makes crypto more accessible to beginners, providing user-friendly interfaces, customer support, and account recovery options.

The tradeoff, however, is security and control. Because centralized exchanges hold large amounts of cryptocurrency in pooled accounts, they become attractive targets for hackers, scammers, and insider threats. When assets are held by a third party, users must trust that platform's security practices—and history has shown that even well-known exchanges are not immune to breaches, freezes, or insolvency.

By contrast, *Decentralized Finance (DeFi)* platforms—such as **Uniswap**, **PancakeSwap**, and **SpookySwap**—operate without intermediaries. Users trade directly from their own wallets using smart contracts, without depositing funds into a centralized account or providing personal identification. This model preserves autonomy and reduces reliance on trusted third parties, but it also shifts all responsibility to the user. There is typically no customer support, no account recovery process, and no recourse if funds are lost due to user error or malicious contracts.

In reality, all cryptocurrency ultimately relies on decentralized blockchain networks—even centralized exchanges depend on these networks to settle transactions. However, the *security model* differs significantly between CeFi and DeFi. Understanding where custody lies, who controls private keys, and what protections exist in each system is critical to making informed decisions and protecting your assets in the crypto ecosystem.

THE IMPORTANCE OF SECURITY IN TRADITIONAL BANKING VS. CRYPTO

In the traditional banking system, security is largely handled by financial institutions on behalf of their customers. Banks invest heavily in both physical and digital safeguards to protect deposits and prevent fraud. Physical security measures include vaults, security guards, motion sensors, and surveillance systems designed to deter theft and unauthorized access. On the digital side, banks rely on passwords, PINs, transaction monitoring, and fraud detection systems to identify suspicious activity.

Perhaps most importantly, traditional banking systems provide insurance and recourse. If fraud occurs, transactions can often be reversed, and customer deposits are typically insured by government-backed programs. These layers of protection mean that, in most cases, the burden of security does not fall entirely on the individual account holder.

Now imagine all of those protections disappearing. This is effectively what decentralized finance looks like.

In the world of cryptocurrency, you are your own bank. There is no customer service department to recover lost or stolen funds, no authority capable of reversing a transaction once it has been confirmed, and no insurance protecting your holdings. If you lose your private key or recovery phrase, access to your assets is lost permanently—regardless of intent or circumstance.

Because of this fundamental difference, securing cryptocurrency is not optional. It is a critical responsibility that comes with participation in decentralized systems. Understanding this shift—from institutional protection to personal accountability—is one of the most important steps in safely using and managing cryptocurrency.

THE REALITY OF CRYPTO SECURITY RISKS

Many cryptocurrency users participate in trading, staking, or mining without fully understanding—or implementing—proper security practices. As a result, countless individuals fall victim to scams, hacks, and phishing attacks, often losing their digital assets permanently. In decentralized systems, these losses are rarely recoverable.

A quick look through crypto communities on platforms like **Reddit**, **Telegram**, or **Discord** reveals an endless stream of cautionary stories: compromised wallets, drained accounts, and irreversible mistakes. While it's easy to believe, "That won't happen to me," the reality is that anyone can be targeted—including experienced traders who become complacent or fall for increasingly sophisticated attacks.

At the same time, responsibility does not rest solely on individual users. The cryptocurrency industry has historically lacked accessible, practical security education, leaving many newcomers unaware of the risks they are assuming. Much like the early days of the internet—when users learned hard lessons about viruses, scams, and privacy—crypto participants are navigating a similar learning curve. Developing strong security habits is not optional; it is a necessary adaptation for anyone choosing to operate in a decentralized financial environment.

HOW THIS BOOK HELPS

The goal of this book is to inform and educate those interested in getting started in cryptocurrency, as well as those already involved but unaware of the unique security risks inherent to the space. Whether you are a beginner or a seasoned trader, taking proactive steps to protect your assets is essential. In crypto, security must be addressed **before** an incident occurs—not after.

Even if you have never experienced a breach or loss, that does not mean you are immune. Strong security practices are not a reaction to failure, but a foundation for long-term participation in the crypto ecosystem. By implementing thoughtful safeguards early, you significantly reduce the likelihood of costly mistakes and help ensure that the assets you've worked to build remain under your control.

CRYPTO SECURITY VS. TRADITIONAL BANKING: EQUIVALENT PROTECTIONS

While decentralized finance lacks the institutional security infrastructure of traditional banks, it does not mean security is absent—it simply shifts responsibility to the individual. In crypto, many personal security practices serve the same functional purpose as the safeguards banks use to protect physical and digital assets.

The comparison below illustrates how common banking protections translate into crypto-specific security measures:

Bank Security Measure	Crypto Equivalent
Bank front door security – physical protection of the building	Using a strong, complex password to prevent unauthorized access
Security guards or receptionists screening entrants	Using a password manager to securely store and manage credentials
Armed guards protecting physical assets	Anti-virus and anti-malware software to defend against hacking attempts
Vault with keys and combinations	A hardware wallet providing offline storage for crypto assets
Video surveillance and alarm systems	Using a VPN and secure network to protect online activity
Bank tellers processing transactions	Hardware wallet verification requiring manual approval

These protections are not direct equivalents, but they highlight an important principle: security in both systems relies on layers. Just as banks do not rely on a single lock or guard, effective crypto security depends on multiple overlapping defenses working together.

FINAL THOUGHTS: YOU ARE YOUR OWN BANK

In traditional finance, banks handle security on your behalf. In cryptocurrency, that responsibility belongs entirely to you.

Without proper security, private keys can be stolen, wallets can be drained in seconds, and losses are often permanent. There is no fraud department to call and no institution capable of restoring access once control is lost.

With proper security practices in place, however, the dynamic changes. You retain control over access to your funds, reduce exposure to common attack vectors, and protect your assets with the same level of care expected of professional investors.

By taking the right precautions today, you can build a secure personal crypto ecosystem—one that keeps your assets safe, accessible, and fully under your control.

CHAPTER 3 - LAYERED SECURITY OR DEFENSE IN DEPTH

UNDERSTANDING SECURITY IN CRYPTOCURRENCY: LAYERED PROTECTION STRATEGIES

Effective cryptocurrency security is not built on a single tool or tactic. Instead, it relies on *layered protection*, often referred to as **defense in depth**. This approach recognizes that no single safeguard is perfect, and that meaningful security comes from multiple overlapping measures working together. If one layer fails, others remain in place to reduce risk and limit potential damage.

When securing cryptocurrency, it is helpful to think in terms of two broad categories of security practices:

1. General Computer and Network Security
These are foundational security principles that apply to all digital environments, including personal computers, home networks, and mobile devices. While many of these concepts may already be familiar, this book examines them specifically through the lens of a crypto holder, trader, or investor. Small weaknesses at the device or network level can have outsized consequences in crypto, where access is direct and losses are often irreversible.

2. Crypto-Specific Security
This category includes security practices unique to cryptocurrency, such as wallet management, private key protection, exchange usage, and transaction verification. Depending on your experience, some of these topics may be familiar, while others may introduce new strategies designed to further safeguard your holdings. These measures address risks that do not exist in traditional finance but are critical in decentralized systems.

Together, these layers form a comprehensive security posture. Understanding how they work—and how they reinforce one another—is essential to protecting digital assets in an environment where responsibility rests entirely with the individual.

THE CORE CONCEPT: DEFENSE-IN-DEPTH (LAYERED SECURITY)

A defense-in-depth strategy—also known as layered security—is a foundational principle for protecting digital assets. Rather than relying on a single safeguard, this approach emphasizes using multiple, overlapping layers of protection. The idea is simple: no single security measure is flawless, but when several are combined, the overall system becomes far more resilient.

Think of security as a maze. An attacker attempting to reach your assets should encounter dead ends, obstacles, and misleading paths at every turn, while only you know the correct route through. Each additional layer increases the time, effort, and resources required to compromise your security. In many cases, attackers will abandon their attempt entirely once they encounter sufficient resistance.

In the context of cryptocurrency, layered security means addressing risks at multiple levels—your accounts, your devices, your network, and your private keys. Examples of effective layered security practices include:

- Using strong, unique passwords for every crypto-related account
- Enabling two-factor authentication (2FA) on exchanges, wallets, and email accounts
- Storing private keys offline using a hardware wallet or secure, encrypted backups
- Using a Virtual Private Network (VPN) to reduce tracking and prevent data interception
- Securing devices with antivirus software, firewalls, disk encryption, and regular updates

Each of these measures alone provides some protection. Together, they create a defensive system that is significantly more difficult to penetrate. This layered approach does not eliminate risk entirely, but it dramatically reduces the likelihood of a single failure leading to total loss—an especially critical consideration in decentralized financial systems.

BALANCING SECURITY AND CONVENIENCE

One of the most common challenges in cybersecurity is balancing security with convenience. In general, the more secure your assets are, the less convenient they may be to access. This tradeoff is unavoidable, especially in cryptocurrency, where security and access are tightly linked.

Strong passwords, for example, are harder to remember but significantly reduce the risk of unauthorized access. Hardware wallets introduce additional steps when accessing funds, but they provide one of the highest levels of protection available. Similarly, two-factor authentication adds a brief delay during login, yet it can stop an attacker even if your password has already been compromised.

The real question is not whether added security creates inconvenience, but whether that inconvenience is acceptable compared to the consequences of a security failure. In practice, the minor effort required to implement strong protections is far preferable to the permanent loss of assets through hacking, scams, or fraud. This balance becomes increasingly important as the value of your cryptocurrency holdings grows.

WHY ONE LAYER IS NOT ENOUGH: THE IMPORTANCE OF MULTIPLE BARRIERS

While tools like two-factor authentication are an excellent security measure, they should never be the only line of defense. Relying on a single protection method creates a single point of failure—something attackers actively look for.

There are several ways even strong security tools can be compromised. A lost or stolen phone can prevent access to authentication codes. SIM swap attacks can allow criminals to intercept text-based 2FA. If an email account is compromised, attackers may also gain access to password resets or account recovery options linked to it.

This is why layered security is essential. When multiple safeguards are in place, the failure of one does not automatically lead to total loss. Instead, additional barriers remain in place to slow, stop, or deter an attacker entirely.

SECURING YOUR COMPUTER AND MOBILE DEVICES

Before focusing on wallets, exchanges, or private keys, it is essential to secure the devices used to access them. Laptops, desktops, mobile phones, and tablets all serve as gateways to cryptocurrency holdings, and each presents its own set of security challenges.

Many security tools can protect multiple device types at once, offering comprehensive coverage across your digital environment. Strengthening these foundational layers is a critical first step, because even the most secure crypto setup can be undermined if the device used to access it is compromised.

UNDERSTANDING MALWARE: THE BIGGEST DIGITAL THREAT

The term *malware*—short for *malicious software*—originally referred to a narrow set of cyber threats. Today, it serves as a broad umbrella term for any software intentionally designed to harm, exploit, or compromise a device, network, or user data. For cryptocurrency users, malware represents one of the most serious digital threats because it often targets access credentials and private keys directly.

Malware can take many forms and may operate silently in the background, making it difficult to detect until damage has already occurred. Common goals of malware include stealing sensitive information such as passwords, banking credentials, and private keys; gaining unauthorized access to a system to allow remote control by an attacker; corrupting or deleting files; and degrading system performance by consuming computing resources.

There are several common categories of malware that users should be aware of:

- **Viruses** are malicious programs that attach themselves to legitimate files and spread when those files are executed.
- **Worms** are self-replicating programs that spread automatically across networks without requiring user interaction.
- **Trojans** disguise themselves as legitimate software but secretly create backdoors that allow unauthorized access.
- **Ransomware** encrypts files or entire systems and demands payment—often in cryptocurrency—in exchange for restoration.
- **Spyware** secretly monitors user activity, including keystrokes, passwords, and browsing behavior.
- **Adware** displays intrusive advertisements and may redirect users to unsafe or malicious websites.

Understanding how malware operates is critical to protecting cryptocurrency holdings. Because crypto assets are controlled directly by users rather than institutions, a single compromised device can lead to irreversible loss. Recognizing malware as a primary threat reinforces the importance of securing devices as a foundational layer in any defense-in-depth strategy.

THE ROLE OF ANTIVIRUS SOFTWARE

Antivirus software has long been a fundamental line of defense against malware. While some antivirus companies have existed for decades and others are newer entrants, the core purpose remains the same: to detect, isolate, and remove malicious software before it can cause harm. For cryptocurrency users, antivirus protection is especially important, as many attacks are designed to compromise devices in order to steal private keys, passwords, or recovery phrases.

One of the most common threats antivirus software is designed to defend against is the computer virus. A virus is a specific type of malware that embeds itself into executable files—such as .exe or .com programs—and spreads when those files are run. Like a biological virus, it often remains dormant until activated, at which point it begins to replicate and infect other programs or areas of the system.

Once active, a virus can cause significant damage. It may steal sensitive information such as login credentials and personal data, corrupt critical system files, or grant attackers remote access to the device—effectively handing over full control without the user's knowledge.

Viruses commonly spread through everyday digital activity. Infected email attachments are a frequent delivery method, particularly when messages appear to come from trusted sources. Downloads from untrusted or unofficial websites can also carry malicious code, as can compromised websites that silently inject malware through embedded scripts. Removable media, such as USB drives or external hard drives, remain another common vector for spreading infected files between systems.

Antivirus software works by monitoring these potential entry points, scanning files for known threats, and using behavioral analysis to identify suspicious activity. While antivirus tools are not a complete security solution on their own, they form a critical foundational layer in a defense-in-depth strategy—especially when combined with safe browsing habits, regular updates, and strong system security practices.

OTHER TYPES OF MALWARE

Beyond traditional computer viruses, several other types of malware pose serious risks to both general computer security and cryptocurrency safety. These threats often operate quietly in the background, making them especially dangerous for users who manage sensitive financial information or private keys.

Worms differ from viruses in that they do not require a host program to spread. Instead, they propagate independently across networks, automatically infecting multiple systems without user interaction. Worms can consume significant system resources, cause network slowdowns, and open the door for additional malware infections.

Trojans, short for Trojan Horses, are among the most deceptive forms of malware. They masquerade as legitimate software—such as games, utilities, email attachments, or even PDF files—while secretly executing malicious commands in the background. Once installed, a Trojan may modify, block, or delete files, steal passwords and private keys, or monitor user activity to collect sensitive information. Because Trojans rely on tricking users into installing them, they are especially common in phishing attacks and fake downloads.

Ransomware is one of the most destructive types of malware. It encrypts files or entire systems and then demands payment—often in cryptocurrency—in exchange for restoring access. If the ransom is not paid, files may remain permanently inaccessible. Even when payment is made, there is no guarantee that attackers will provide decryption keys, making ransomware particularly devastating.

Spyware is designed to secretly monitor user behavior. It may record keystrokes through keyloggers to capture passwords and recovery phrases, track browsing history, or collect financial data such as credit card information. Because spyware often runs silently, users may remain unaware of the compromise until funds are stolen or accounts are drained.

Adware is sometimes viewed as a nuisance rather than a serious threat, but it can still pose meaningful risks. Adware bombards users with intrusive advertisements, redirects browsers to unsafe websites, and degrades system performance. In more aggressive cases, adware may include spyware components, allowing it to track activity or collect data without the user's consent.

Understanding these different forms of malware reinforces why device security is a critical foundation for cryptocurrency protection. Any malware capable of monitoring activity, intercepting data, or controlling a device can ultimately compromise access to digital assets.

KEY TAKEAWAYS FOR PROTECTING YOUR DEVICES

Protecting your devices is one of the most important steps in safeguarding your cryptocurrency holdings, as compromised hardware often leads directly to compromised assets. Using a reputable antivirus program provides a foundational layer of defense by detecting and removing known threats before they can cause damage. Keeping that software up to date—preferably through automatic updates—is essential, as new threats emerge constantly.

Safe browsing habits are equally important. Avoid clicking on suspicious links, downloading unverified files, or opening unexpected attachments, even if they appear to come from familiar sources. Email attachments in particular should be treated with caution and scanned before opening, as they remain one of the most common delivery methods for malware.

Additional tools can further reduce risk. Ad-blockers and browser security extensions help prevent malicious scripts, deceptive advertisements, and unsafe redirects from loading in the first place. These tools act as another barrier, stopping threats before they ever reach your system.

Finally, regular backups of critical files are an essential safeguard—especially against ransomware. While backups do not prevent an attack, they ensure that important data can be restored without paying a ransom or suffering permanent loss.

Together, these practices form a strong baseline for device security. In the next section, we will explore specific antivirus solutions and how to choose the best protection for your particular setup.

TYPES OF COMPUTER VIRUSES AND THEIR EFFECTS

Computer viruses are one of the oldest and most persistent forms of malware, and they remain a significant threat to system security today. At a basic level, a virus is malicious code designed to attach itself to legitimate files or system components, allowing it to spread, execute unwanted actions, or disrupt normal operations—often without the user realizing it until damage has already occurred.

One of the most common forms is the file-infecting virus. These viruses attach themselves to executable programs, such as .exe or .com files. When an infected program is launched, the virus activates and may overwrite files, corrupt data, or spread to other programs on the system. Because the malicious code hides inside otherwise legitimate software, users often trigger these infections unknowingly.

Macro viruses take advantage of scripting features built into applications like Microsoft Word and Excel. They typically spread through infected documents shared via email or file-sharing platforms. When the document is opened, the malicious macro executes automatically, potentially stealing information, altering files, or downloading additional malware in the background.

Another common threat is the browser hijacker, which focuses less on system destruction and more on manipulation and data collection. These infections alter browser settings to redirect users to malicious websites, inject unwanted advertisements, or modify search results. Beyond being disruptive, browser hijackers pose serious privacy risks by tracking browsing behavior and harvesting personal data.

Web scripting viruses operate at the intersection of browsers and websites. These attacks inject malicious code into legitimate web pages, often without the site owner's knowledge. When users visit a compromised site, the virus can steal cookies, capture login credentials, or gain access to linked accounts such as email or social media—sometimes allowing attackers to impersonate the user entirely.

Some viruses target systems at a deeper level. A boot sector virus, for example, infects the portion of a computer responsible for starting the operating system. While once commonly spread through floppy disks, modern versions are more likely to propagate through USB drives or external storage. Because these viruses load into memory before the operating system fully initializes, they can render systems unbootable and difficult to recover without advanced tools.

More sophisticated threats include polymorphic viruses, which are specifically designed to evade detection. Each time they execute, they alter or encrypt their own code while preserving functionality. This constant mutation makes them far more difficult for traditional antivirus software to identify, often requiring advanced behavioral analysis to detect and remove.

A resident virus embeds itself directly into system memory, allowing it to remain active even after the original infected file is deleted. These viruses can slow performance, corrupt files, crash applications, and persist through system reboots, making them particularly challenging to eliminate.

Finally, multipartite viruses represent one of the most dangerous categories. Rather than targeting a single system component, they attack multiple areas simultaneously—such as files, memory, and the boot sector. This layered infection strategy makes containment and cleanup far more complex and significantly increases the potential for widespread system damage if not addressed quickly.

PREVENTING VIRUS INFECTIONS: KEY SECURITY MEASURES

Preventing computer viruses, much like preventing biological ones, requires proactive and consistent protective measures. Many infections occur not because of sophisticated attacks, but due to everyday behaviors such as opening unverified attachments or clicking unknown links.

Using a reputable antivirus program—and keeping it regularly updated—provides a foundational layer of defense by detecting and neutralizing known threats. Routine system scans help identify infections early, before they have time to spread or cause lasting damage.

Equally important is keeping your operating system and software up to date. Security updates and patches fix known vulnerabilities that viruses frequently exploit. Firewalls add another protective barrier by blocking unauthorized access attempts and limiting how malware can communicate once inside a system.

By understanding how viruses operate and maintaining strong digital hygiene, users can significantly reduce their exposure to malware threats. These practices form an essential layer of defense in any comprehensive cryptocurrency security strategy.

PROTECTING YOURSELF FROM MALWARE AND CYBER THREATS

While antivirus programs are an essential layer of protection, no security software can provide complete immunity. User awareness and daily habits play an equally critical role in preventing malware infections. Many successful attacks rely not on technical sophistication, but on convincing users to take a single unsafe action.

One of the most common entry points for malware is suspicious links. Phishing attacks frequently arrive via email, social media, or text messages and are designed to appear legitimate or urgent. If you receive an unexpected link—especially one prompting immediate action—it should not be clicked unless its authenticity can be independently verified.

Email attachments also require caution. Even messages that appear to come from trusted contacts can be compromised, as attackers often impersonate friends, coworkers, or service providers. Scanning attachments with antivirus software before opening them helps prevent malicious files from executing silently in the background.

Another frequent threat comes from malicious advertisements, often referred to as malvertising. Clicking on pop-up ads or banner advertisements—particularly on untrusted websites—can lead to malware infections. Using pop-up blockers and avoiding ad interactions altogether significantly reduces this risk.

Unexpected website redirects are another warning sign. If a link suddenly takes you to a different site than anticipated, do not enter personal information, download files, or click additional links. Close the browser tab immediately and avoid further interaction.

Finally, maintaining a reliable antivirus program is essential. Keeping antivirus software updated ensures protection against newly discovered threats. Enabling real-time scanning and scheduling regular system scans helps detect malicious activity early, before it can compromise sensitive data or access cryptocurrency-related information.

Together, these practices form a strong behavioral layer in your defense-in-depth strategy. By combining technical tools with informed decision-making, you significantly reduce the likelihood that malware will gain a foothold on your devices.

CHOOSING THE RIGHT ANTIVIRUS SOFTWARE

While antivirus software is a necessary component of device security, not all solutions perform equally. Some programs can noticeably slow down a system due to constant background scanning or poorly optimized processes. Selecting the right antivirus software requires balancing strong protection with minimal impact on everyday performance.

One important factor to consider is performance impact. Independent reviews and testing organizations often measure how much a security program affects system speed during normal use. Choosing software that provides effective protection without significantly degrading performance helps ensure that security does not become a burden.

Real-time scanning is another essential feature. A quality antivirus solution should actively monitor files, downloads, and network activity as they occur, rather than relying solely on manual scans. This allows threats to be detected and blocked before they can execute.

Scheduled scanning adds an additional layer of protection without disrupting productivity. Automatic scans can be configured to run during periods when the computer is not in use, such as overnight, ensuring regular system checks without affecting daily tasks.

Detection accuracy is equally critical. Antivirus programs should demonstrate high malware detection rates based on independent testing, as missed threats can be far more costly than minor performance slowdowns.

Some antivirus solutions also include additional security tools, such as built-in firewalls, password managers, or virtual private networks (VPNs). While these features are not always necessary, they can enhance overall security when properly implemented and used as part of a broader defense-in-depth strategy.

OTHER FORMS OF MALWARE TO WATCH OUT FOR

Beyond traditional computer viruses, a wide range of other malicious programs pose serious security threats—often in more subtle and dangerous ways. These forms of malware are frequently designed not to damage systems outright, but to quietly monitor behavior, extract sensitive information, or extort victims over time.

Spyware operates silently in the background, often without any visible signs of infection. Its primary purpose is surveillance. Spyware can capture login credentials, financial information, and other personal data, transmitting it to a remote attacker without the user's knowledge. One particularly invasive form of spyware is the keylogger, which records every keystroke made on a device. This allows attackers to capture passwords, private messages, and even recovery phrases as they are typed—making spyware especially dangerous for anyone managing financial accounts or cryptocurrency wallets.

Adware is sometimes dismissed as a nuisance rather than a threat, but it can still pose meaningful security risks. Adware tracks browsing habits in order to display targeted advertisements, often redirecting users to questionable or malicious websites. In more aggressive forms, adware may include Trojan components that collect personal data or open the door to additional malware infections. Over time, heavy adware presence can significantly slow system performance and make normal browsing frustrating or unsafe.

Among the most disruptive threats is ransomware, which is designed to lock users out of their own data. Once activated, ransomware encrypts important files, rendering them inaccessible until a ransom is paid. These attacks commonly originate from phishing emails, fake software downloads, or compromised websites. Payment is almost always demanded in cryptocurrency, allowing attackers to operate across borders while minimizing traceability. Even when a ransom is paid, there is no guarantee that access will be restored, making prevention and early detection absolutely critical.

PHISHING ATTACKS: A GROWING CYBERSECURITY THREAT

WHAT IS PHISHING?

Phishing is a type of social engineering attack designed to trick individuals into revealing sensitive information by impersonating legitimate and trusted sources. These attacks commonly appear as emails, text messages, phone calls, or fake websites that mimic banks, employers, service providers, or even friends and colleagues. The goal is to create a sense of urgency or legitimacy—prompting the recipient to click a link, download an attachment, or enter confidential information such as passwords, recovery phrases, or financial details. Unlike many technical attacks that exploit software vulnerabilities, phishing exploits human behavior, making it one of the most effective and dangerous threats in both traditional finance and cryptocurrency environments.

WEBSITE PHISHING: A THREAT TO CRYPTO USERS

Website phishing is one of the most dangerous threats facing cryptocurrency users because it targets trust rather than technology. In these attacks, criminals create fake versions of legitimate cryptocurrency exchanges, wallet interfaces, or popular DeFi platforms that are nearly indistinguishable from the real ones. Logos, layouts, color schemes, and even domain names are carefully replicated, often differing by only a single character or subtle spelling variation that is easy to miss.

These fraudulent sites are commonly distributed through phishing emails, search engine advertisements, social media links, or fake support messages claiming there is an urgent issue with an account. Once a user arrives at the counterfeit website, they are prompted to log in, connect a wallet, or enter sensitive information such as passwords, recovery phrases, or private keys. The moment this information is submitted, it is captured by the attacker—often resulting in immediate and irreversible loss of funds.

In cryptocurrency systems, there is no safety net once this occurs. Unlike traditional banking, transactions cannot be reversed, and there is no central authority capable of restoring stolen assets. This is what makes website phishing particularly devastating in crypto: a single mistake can permanently transfer control of assets to an attacker.

Recognizing that even experienced users can be fooled is essential. Awareness, not confidence alone, is what protects users in decentralized financial systems.

HOW TO SPOT WEBSITE PHISHING SCAMS:

Learning to identify phishing websites is a critical defensive skill. One of the most important checks is the website address itself. Phishing sites often use subtly altered domain names, such as added characters, misspellings, or unfamiliar extensions designed to look legitimate at a glance.

It is also important to avoid clicking links from unsolicited emails, messages, or social media posts. Instead, manually typing known website addresses into your browser or using saved bookmarks significantly reduces the risk of being redirected to a fraudulent site.

Finally, pay attention to basic security indicators. Legitimate websites should use encrypted connections, typically indicated by "HTTPS" in the address bar and, in many browsers, a padlock icon. While these indicators alone do not guarantee a site is safe, their absence is a strong warning sign.

FINAL THOUGHTS: THE IMPORTANCE OF VIGILANCE

Cybercriminals are constantly developing new tactics to bypass security measures, which means no single tool or setting can provide complete protection. The most effective defense is a combination of reliable security software and informed, deliberate user behavior.

Thinking before you click is one of the simplest yet most powerful habits you can develop. Always verify links, senders, and attachments before interacting with them—especially when messages create urgency or request sensitive information. Maintaining strong security tools is equally important. Antivirus software, firewalls, and ad blockers should remain enabled and up to date to reduce exposure to common attack vectors.

Staying informed is another critical layer of defense. Cyber threats evolve continuously, and understanding new attack methods helps prevent complacency. Regularly updating your knowledge ensures that yesterday's safe habits do not become today's vulnerabilities.

Finally, enabling multi-factor authentication adds a crucial barrier against unauthorized access. Even if a password is compromised, additional verification steps can prevent attackers from gaining control of accounts or assets.

By combining these practices, you significantly reduce the likelihood of falling victim to cyber threats and strengthen your overall security posture—protecting both your devices and your cryptocurrency holdings.

Note: *None of the products discussed in this book are endorsements or recommendations. They are presented solely for educational purposes to illustrate various tools, concepts, and approaches to network and cybersecurity. Readers are encouraged to conduct their own research and due diligence to determine which products or solutions best meet their individual needs.*

PROTECTING YOUR SYSTEM: ESSENTIAL SECURITY MEASURES

Now that we've covered the general threats facing computers and mobile devices, it's time to focus on practical remedies and preventive measures. Effective security is not about applying every possible tool, but about selecting the right protections for your specific environment and using them consistently.

Important disclaimer: Every situation is different, and each operating system—whether Windows, macOS, Android, or iOS—offers its own security features and limitations. The strategies discussed in this section are intended to provide general guidance, not universal solutions. Readers are encouraged to do their own research and choose security measures that best align with their devices, budget, and personal needs.

1. Keeping Your System Up to Date

Regular software updates are one of the simplest—and most effective—ways to protect your devices. Updates frequently patch known security vulnerabilities that attackers actively exploit. When operating systems, browsers, or applications remain outdated, those weaknesses can be used to steal data, install malware, or compromise access to cryptocurrency wallets and accounts.

Keeping your operating system fully updated—whether you use Windows, macOS, Android, or iOS—should be considered a baseline security practice. The same applies to web browsers such as Chrome, Firefox, Edge, or Safari, which are common attack surfaces due to their constant interaction with the internet. Frequently used applications, including messaging and communication tools, should also be kept current, as they are often targeted through malicious links or attachments.

Security tools require updates as well. Antivirus software, firewalls, VPNs, and cryptocurrency wallets all rely on regular updates to remain effective against newly discovered threats. Enabling automatic updates wherever possible reduces the risk of forgetting critical patches and minimizes reliance on manual maintenance—one of the most common points of failure in personal security.

2. Choosing the Right Antivirus Software

Antivirus software plays a critical role in detecting, preventing, and removing malware. These programs provide real-time protection by monitoring files, downloads, and system behavior for signs of malicious activity. While built-in protections—such as Windows Defender—are effective for many users, others prefer dedicated antivirus solutions for additional layers of security and advanced features.

Several well-established antivirus providers offer cross-platform protection and robust threat detection, including McAfee, Norton, Kaspersky, Bitdefender, and Malwarebytes. While features vary between providers, most offer core protections such as malware detection, real-time scanning, and system monitoring.

Users with multiple devices should pay particular attention to platform compatibility. Choosing a security package that protects all devices—such as a Windows laptop and a mobile phone—under a single subscription can simplify management and ensure consistent protection across your digital environment.

3. Research Before You Buy: Finding the Best Security Package

Security needs vary from person to person, and no single solution is ideal for everyone. Before selecting an antivirus or security suite, it's important to consider whether the software supports all of your devices and operating systems. Performance impact is another key factor—strong protection should not significantly slow down everyday use.

Feature sets also matter. Some security packages include additional tools such as password managers, firewalls, VPNs, or enhanced real-time scanning. While these features can add value, they should complement—not replace—good security habits. Pricing models and subscription terms are also worth reviewing, particularly for users managing multiple devices.

It's important to remember that no single program can provide complete protection. A layered security approach—combining regular updates, strong passwords, antivirus protection, firewalls, and informed user behavior—offers the most reliable defense.

FINAL TAKEAWAY: BE PROACTIVE, STAY SECURE

Cybersecurity is an ongoing process, not a one-time setup. Threats evolve, software changes, and attack methods adapt over time. By regularly updating your systems, using reliable security tools, and staying informed about emerging risks, you significantly reduce the likelihood of malware infections, hacking attempts, and cryptocurrency theft.

In a decentralized financial environment, security is not something that can be delegated or ignored. Proactive habits and consistent vigilance are what keep digital assets safe—today and in the long term.

CHAPTER 4 - BASIC COMPUTER SECURITY

WEB BROWSERS: A CRITICAL YET OVERLOOKED SECURITY CONSIDERATION

One of the most overlooked security vulnerabilities in personal computing is the web browser. While many users focus on protecting their devices with antivirus software and firewalls, they often underestimate the role browsers play as both a primary attack surface and a major privacy risk.

Web browsers are the gateway to nearly everything online—email, exchanges, wallets, social media, and financial services. As a result, they are a frequent target for malware, phishing attempts, malicious scripts, and tracking technologies. Some widely used browsers lack strong privacy protections by default, while others actively collect and report user data to parent companies for advertising, analytics, or behavioral profiling.

While this type of data collection may not directly compromise cryptocurrency holdings, it can significantly increase exposure to phishing attacks, targeted scams, tracking, and data leaks. Over time, these risks compound—making it easier for attackers to profile users, impersonate trusted services, or deliver highly convincing malicious content.

When securing digital assets, it's important to remember two fundamental principles:

- You are your own bank.
- Your web browser plays a critical role in protecting your digital identity.

UNDERSTANDING BROWSER SECURITY AND THE CHROMIUM ECOSYSTEM

Most modern web browsers are built on Chromium, an open-source browser project originally developed and maintained by Google. Open-source software offers clear advantages, including transparency and the ability for independent researchers to audit code for security flaws. However, not all Chromium-based browsers offer the same level of privacy or user protection.

Several popular browsers are built on Chromium, including:

- Google Chrome – The most widely used browser, but one that collects extensive user data by default.
- Microsoft Edge – A Chromium-based browser with improved performance and some privacy controls, though still closely integrated with Microsoft services.
- Opera – Includes built-in ad blocking and a free VPN feature, but has mixed reviews regarding privacy practices.
- Brave – A privacy-focused browser that removes Google tracking from Chromium and blocks ads and trackers by default.

Because these browsers share a common underlying framework, they often receive similar security updates and performance improvements. However, their privacy policies, default settings, and data-collection practices can differ significantly. Understanding these differences is critical when choosing a browser that aligns with a security-first mindset—especially for users managing financial accounts or cryptocurrency assets.

SOME WEB BROWSERS BETTER SUITED FOR SECURITY AND PRIVACY

If online security and privacy are a priority, it may be worth considering alternatives to mainstream browsers such as Chrome or Edge. While popular browsers offer convenience and broad compatibility, they often prioritize data collection, analytics, and advertising integration over privacy by default.

Browsers that focus on security and privacy typically reduce tracking, limit data collection, and offer stronger default protections against malicious scripts, fingerprinting, and third-party surveillance. These features are especially relevant for cryptocurrency users, as targeted advertising, behavioral profiling, and browser-based tracking can increase exposure to phishing attempts and social engineering attacks.

It is important to note that no browser is inherently "secure" out of the box. Security depends on how a browser is configured, how it is used, and how well it fits into a broader defense-in-depth strategy. That said, some browsers are designed with privacy and user control as core principles rather than optional settings.

Examples of browsers that place a stronger emphasis on privacy and security include:

- Brave – Built on Chromium but removes Google tracking, blocks ads and trackers by default, and offers strong privacy protections with minimal configuration.

- Mozilla Firefox – An open-source browser with extensive privacy controls, strong extension support, and a long-standing focus on user autonomy when properly configured.
- Tor Browser – Designed for maximum anonymity by routing traffic through the Tor network, though at the cost of speed and convenience.
- LibreWolf – A hardened, privacy-focused fork of Firefox that removes telemetry and applies strict default security settings.

Each of these browsers involves tradeoffs between usability, compatibility, speed, and privacy. The best choice depends on how you use the internet, what devices you rely on, and how much friction you are willing to accept in exchange for stronger protections.

In the sections that follow, we'll explore how browser configuration, extensions, and usage habits can matter just as much as the browser itself—especially when accessing financial accounts or managing cryptocurrency assets.

Note: *None of the products discussed in this book are endorsements or recommendations. They are presented solely for educational purposes to illustrate various tools, concepts, and approaches to network and cybersecurity. Readers are encouraged to conduct their own research and due diligence to determine which products or solutions best meet their individual needs.*

1. **Brave – A Privacy-Focused Chromium Browser**

✅ Blocks ads, trackers, and browser fingerprinting by default
✅ Removes Google tracking components from Chromium
✅ Offers optional Tor private windows for enhanced anonymity
✅ Fast, familiar, and compatible with most modern websites

WHY CHOOSE BRAVE?
Brave is a Chromium-based browser designed to reduce tracking and advertising while maintaining performance and usability. Unlike Google Chrome, it does not collect browsing history or behavioral data by default. This makes it a strong option for users who want the familiarity and compatibility of Chromium without extensive data collection.

Brave is particularly well suited for everyday secure browsing, including accessing exchanges and crypto-related websites, with minimal configuration required.

2. **Mozilla Firefox – A Flexible, Privacy-Respecting Alternative**

✅ Enhanced Tracking Protection blocks many trackers automatically
✅ Reduces browser fingerprinting and cross-site tracking
✅ Highly customizable with advanced privacy controls
✅ Open-source and independent from Chromium

WHY CHOOSE FIREFOX?
Firefox remains one of the strongest non-Chromium browser options for privacy-conscious users. Its open-source development model and extensive configuration options allow users to tailor privacy and security settings to their needs.

While Firefox may not always receive Chromium-specific security features as quickly, it excels in user control and transparency. For users who value configurability and independence from large advertising ecosystems, Firefox is a strong choice—especially when properly hardened.

3. **LibreWolf – A Hardened Firefox for Maximum Privacy**

✅ Removes telemetry, analytics, and background connections
✅ Blocks trackers, fingerprinting, and third-party cookies by default
✅ Enforces strong security settings out of the box
✅ Community-driven and privacy-first by design

WHY CHOOSE LIBREWOLF?

LibreWolf is a hardened fork of Firefox designed for users who want aggressive privacy protections without manual configuration. It disables data collection features and applies strict security defaults that significantly reduce tracking and data leakage.

Because of these protections, some websites may not function correctly without adjusting permissions. LibreWolf is best suited for sensitive activities such as financial management or cryptocurrency access, particularly for users willing to trade some convenience for stronger default security.

4. **Tor Browser – Maximum Anonymity with Tradeoffs**

- ✅ Routes traffic through multiple encrypted relays
- ✅ Prevents IP address and location identification
- ✅ Strong protections against fingerprinting and tracking

WHY CHOOSE TOR?

Tor Browser is designed for anonymity rather than convenience. By routing traffic through the Tor network, it makes tracking users extremely difficult. This level of protection is valuable for highly sensitive activities, but it comes with limitations.

Common drawbacks include slower browsing speeds, frequent CAPTCHAs, and websites that may not function properly. For most users, Tor is best reserved for specific situations where anonymity is essential rather than for everyday browsing.

Best Practice: Many security-conscious users adopt a browser compartmentalization strategy—using Brave or Firefox for daily browsing and Tor or LibreWolf for sensitive financial or cryptocurrency-related tasks.

FINAL THOUGHTS: SECURING YOUR BROWSER FOR CRYPTO SAFETY

Regardless of which browser you choose, additional security measures are essential for protecting both your cryptocurrency assets and your online privacy. Privacy-focused browsers such as Brave, Mozilla Firefox, or Tor Browser provide a stronger starting point than mainstream alternatives, but configuration and user behavior ultimately determine how effective those tools are.

Disabling JavaScript on unfamiliar or untrusted websites can help prevent malicious scripts from executing silently in the background. While JavaScript is required for many modern sites to function, selectively limiting it reduces exposure to drive-by attacks and exploit kits. Regularly clearing cookies and browsing history also helps minimize tracking and reduces the amount of data that can be leveraged for profiling or targeted phishing.

Credential management deserves special attention. Browser-based password managers are convenient, but they increase risk if the browser itself is compromised. Using a dedicated password manager isolates credentials from browsing activity and provides an additional layer of protection. Finally, always verify website URLs carefully before entering login information—especially when accessing cryptocurrency exchanges, wallets, or financial platforms. A single mistyped or spoofed domain can result in immediate and irreversible loss.

By choosing a browser that aligns with your security goals and reinforcing it with deliberate practices, you significantly reduce your exposure to online threats. In cryptocurrency, where users are fully responsible for their own security, browser hygiene is not optional—it is a critical part of protecting both personal data and digital assets.

ENHANCING PRIVACY WITH ENCRYPTED EMAIL SERVICES

Email is one of the most widely used digital communication tools, yet it was never designed with privacy or security as a priority. Despite this, email is commonly used for account registration, password recovery, transaction alerts, and customer support—including for financial services and cryptocurrency platforms.

Many users rely on mainstream providers such as **Gmail**, **Yahoo Mail**, or **Outlook**. While convenient and reliable, these services often scan email content, collect metadata, and analyze user behavior for advertising, analytics, or compliance purposes. Even when message content is encrypted in transit, providers themselves may still have access to the data stored on their servers.

For cryptocurrency users, this creates a meaningful risk. Email accounts are frequently targeted in phishing attacks, account takeovers, and recovery-based exploits. If an attacker gains access to your email, they may be able to reset passwords, intercept security alerts, or impersonate you when contacting exchanges or service providers.

Encrypted email services address these risks by using end-to-end encryption, meaning messages are encrypted on the sender's device and can only be decrypted by the intended recipient. In this model, even the email provider cannot read the contents of messages. This significantly reduces the risk of data exposure, surveillance, or unauthorized access—especially when communicating about sensitive financial matters.

While email will likely remain an essential tool, using an encrypted email provider for cryptocurrency-related accounts, financial communications, and account recovery can dramatically improve your overall security posture. In a system where personal responsibility is paramount, protecting email access is not just a privacy choice—it is a foundational layer of crypto security.

WHY STANDARD EMAIL IS INSECURE

Standard email was designed for convenience and interoperability—not security. As a result, most traditional email systems lack protections that are essential when communicating about sensitive topics such as financial accounts or cryptocurrency holdings.

One major weakness is the absence of built-in end-to-end encryption. While many providers encrypt emails while they are in transit between servers, the messages are often stored in a readable form once they reach the provider. This means the service itself—and potentially anyone who gains access to its systems—can read message contents.

In addition, many free email services actively scan emails and collect metadata. This scanning may be used for advertising, analytics, spam filtering, or regulatory compliance, but it still represents a loss of privacy and an expanded attack surface. Even when message content is not explicitly read by humans, automated analysis can expose patterns about user behavior, interests, and relationships.

Email accounts are also prime targets for attackers. Because email is commonly used for password resets, account recovery, and security alerts, gaining access to an email inbox can allow an attacker to compromise multiple connected accounts. Phishing attacks, credential theft, and account takeovers frequently begin with email exploitation, making it one of the weakest links in personal security.

While encrypted email services address many of these issues, they are not without limitations. Understanding both the strengths and constraints of encrypted email is essential, and we will explore those considerations in the sections that follow.

BEST ENCRYPTED EMAIL PROVIDERS FOR SECURITY & PRIVACY

For cryptocurrency users, email security is not optional. Email accounts are commonly used for exchange logins, password recovery, transaction alerts, and customer support communication. If an attacker gains access to your email, they may be able to compromise multiple connected accounts—even without direct access to your wallets.

Encrypted email services address many of the weaknesses found in standard email by implementing end-to-end encryption (E2EE), ensuring that messages can only be read by the intended sender and recipient. While no solution is perfect, using encrypted email for crypto-related communication significantly strengthens your overall security posture.

Below are several encrypted email solutions commonly used by privacy-conscious individuals. These are discussed for educational purposes, not as endorsements.

Note: *None of the products discussed in this book are endorsements or recommendations. They are presented solely for educational purposes to illustrate various tools, concepts, and approaches to network and cybersecurity. Readers are encouraged to conduct their own research and due diligence to determine which products or solutions best meet their individual needs.*

PROTONMAIL – A LEADING SECURE EMAIL PLATFORM

Based in Switzerland, ProtonMail benefits from strong privacy protections under Swiss law. It uses OpenPGP-based end-to-end encryption to secure email content and related data. ProtonMail allows encrypted communication both between Proton users and with external recipients through password-protected messages.

ProtonMail offers mobile and desktop applications, as well as a web interface. However, it is important to understand that viewing ProtonMail through a web browser alone provides HTTPS-level encryption rather than full end-to-end encryption. True E2EE with ProtonMail requires using an email client in conjunction with ProtonMail Bridge, which integrates with applications such as Outlook, Thunderbird, or Apple Mail.

ProtonMail is well suited for users who prioritize privacy and are comfortable using dedicated tools to maximize encryption.

TUTANOTA – BUILT-IN ENCRYPTION WITH MINIMAL CONFIGURATION

Tutanota is a Germany-based encrypted email provider that uses its own encryption system rather than OpenPGP. Emails are automatically encrypted without requiring third-party tools or bridges, making it one of the most user-friendly encrypted email options available.

Tutanota encrypts email content, subject lines, and contacts, and allows users to send encrypted messages to non-Tutanota recipients using a shared password. The service offers mobile and desktop apps, free and paid plans, and does not rely on advertising or user tracking.

Tutanota is an excellent choice for users who want strong encryption with minimal setup and complexity.

SECUREMYEMAIL – ENCRYPTION WITHOUT CHANGING PROVIDERS

SecureMyEmail takes a different approach by adding end-to-end encryption to existing email accounts such as Gmail, Yahoo, or Outlook. Instead of creating a new email address, users encrypt messages on top of their current provider using PGP-based encryption.

This solution allows encrypted communication with any recipient and avoids the inconvenience of switching email addresses. SecureMyEmail is a paid service and is best suited for users who want encryption while continuing to use their existing email ecosystem.

While this approach improves message security, users should remember that metadata and account access remain dependent on the underlying email provider.

ZOHO MAIL – BUSINESS-ORIENTED SECURE EMAIL

Zoho Mail is designed primarily for business users and offers secure email hosting with optional encryption features. It provides encrypted data storage, collaboration tools, and professional email hosting with custom domains.

Unlike privacy-first encrypted providers, Zoho Mail can be subject to legal subpoenas depending on jurisdiction. While it follows established legal protocols, it does not provide the same anonymity or zero-knowledge encryption model as ProtonMail or Tutanota.

Zoho Mail is best suited for users who prioritize business functionality and structured security over maximum privacy or anonymity.

CHOOSING THE RIGHT ENCRYPTED EMAIL PROVIDER

There is no single "best" encrypted email solution for everyone. The right choice depends on your priorities:

If maximum privacy and strong jurisdictional protections matter most, ProtonMail or Tutanota are strong options.

If you want encryption without changing your existing email address, SecureMyEmail may be preferable.

If you need business-grade email with security controls, Zoho Mail may be appropriate.

Regardless of provider, encrypted email should be viewed as one layer of a broader defense-in-depth strategy—not a standalone solution.

FINAL THOUGHTS: WHY ENCRYPTED EMAIL MATTERS FOR CRYPTO USERS

Email remains one of the most common attack vectors used to compromise cryptocurrency accounts. Using encrypted email reduces the risk of phishing, limits data exposure, and makes it more difficult for attackers to profile or impersonate you.

For crypto users, encrypted email helps protect:

- Account recovery communications
- Exchange and wallet alerts
- Sensitive financial discussions
- Personal identity and metadata

While encrypted email does not eliminate all risks, it significantly raises the barrier for attackers. In a decentralized financial environment where responsibility rests entirely with the user, securing email access is a logical and necessary step toward protecting digital assets and online identity.

CHAPTER 5 - PASSWORD VAULTS AND 2FA

PASSWORD VAULTS: ENHANCING SECURITY AND CONVENIENCE

A password vault—commonly referred to as a password manager—is an encrypted storage system designed to securely hold usernames, passwords, credit card information, and other sensitive data. Rather than relying on memory or insecure storage methods, password vaults centralize credential management behind strong encryption.

These tools address one of the most common security failures in digital life: password reuse. When users rely on the same or similar passwords across multiple sites, a single breach can cascade into widespread account compromise. Password vaults solve this problem by generating and storing unique, complex passwords for every service—without sacrificing usability.

By bridging the gap between security and convenience, password vaults play a critical role in protecting financial accounts, email access, and cryptocurrency platforms.

WHY NOT USE YOUR BROWSER'S BUILT-IN PASSWORD MANAGER?

Most modern web browsers offer to store passwords automatically, often prompting users with messages like:

"Would you like your browser to remember this password?"

While convenient, browser-based password storage introduces meaningful risks—especially for users managing sensitive financial or cryptocurrency accounts.

Browser password managers often generate weaker or less configurable passwords compared to dedicated vaults. More importantly, credentials stored directly in a browser are only as secure as the browser itself. If a browser is compromised through malware, malicious extensions, or phishing attacks, attackers may gain access to all saved credentials at once.

In addition, browser-based password storage can suffer from limited cross-platform security. While syncing across devices is possible, it often relies on cloud accounts that may become another point of failure if compromised.

For cryptocurrency users, this risk is particularly serious. Gaining access to browser-stored passwords can grant attackers entry to exchanges, email accounts, wallets, and recovery mechanisms—often with no way to reverse the damage.

BENEFITS OF USING A PASSWORD VAULT

Dedicated password vaults are designed with security as their primary function, offering protections that browsers simply do not provide.

They use strong, industry-standard encryption to protect stored credentials, ensuring that even if the vault provider were compromised, the data itself would remain unreadable. Each account can have a unique, randomly generated password, preventing credential-stuffing attacks if one service is breached.

Password vaults also reduce exposure to keylogging malware by minimizing manual typing. Auto-fill and auto-login features allow credentials to be entered securely without passing through the keyboard. Many vaults also support multiple operating systems—Windows, macOS, Linux, iOS, and Android—making secure access consistent across devices.

Some password managers monitor known data breaches and alert users if stored credentials have been exposed, allowing proactive password changes before damage occurs. All of this is protected behind a single master password, meaning users only need to remember one strong credential instead of dozens.

By using a password vault, you significantly improve security while simplifying daily digital life. For cryptocurrency holders—where account compromise can result in irreversible loss—this balance of protection and convenience is not just helpful, it is essential.

SOME PASSWORD VAULTS WELL SUITED FOR SECURITY AND PRIVACY

For cryptocurrency users, password vaults are not just a convenience—they are a foundational security tool. Exchanges, wallets, email accounts, and recovery mechanisms all depend on credentials that must remain isolated, unique, and protected. A dedicated password vault dramatically reduces the risk of credential reuse, phishing fallout, and account takeover.

The following password managers are discussed for educational purposes only, not as endorsements. Each represents a different approach to security, privacy, and usability.

Note: *None of the products discussed in this book are endorsements or recommendations. They are presented solely for educational purposes to illustrate various tools, concepts, and approaches to network and cybersecurity. Readers are encouraged to conduct their own research and due diligence to determine which products or solutions best meet their individual needs.*

1PASSWORD

1Password is a polished, feature-rich password manager designed for both individuals and organizations. It uses end-to-end encryption and offers strong cross-platform support across Windows, macOS, Linux, iOS, Android, and browser extensions.

One of its standout features is Travel Mode, which allows users to temporarily remove selected vaults from their devices when crossing borders, restoring access later. This can reduce exposure during device inspections or travel-related risks. 1Password also includes alerts for weak or compromised passwords and supports multiple vaults and user profiles.

1Password is best suited for users who want premium features, strong usability, and travel-conscious security in a managed ecosystem.

BITWARDEN

Bitwarden is a widely respected open-source password manager that emphasizes transparency and user control. Its zero-knowledge encryption model ensures that even Bitwarden itself cannot access stored credentials. The codebase is publicly auditable, and the platform has undergone multiple third-party security reviews.

Bitwarden supports optional cloud syncing as well as self-hosting for users who want complete control over their vault infrastructure. A free tier is available, with paid plans offering advanced features such as enhanced security reporting and emergency access.

Bitwarden is an excellent choice for users who value open-source security, flexibility, and independence from proprietary ecosystems.

KEEPASS

KeePass takes a fundamentally different approach by operating entirely offline. Password databases are stored locally, with no built-in cloud synchronization. This dramatically reduces exposure to cloud-based breaches but places full responsibility for backups and file management on the user.

KeePass is highly customizable through community-developed plugins and supports strong encryption standards. While the core application is Windows-based, ports and compatible clients exist for macOS, Linux, Android, and iOS.

KeePass is best suited for technically confident users who want maximum control and are comfortable managing backups and synchronization manually.

NORDPASS

NordPass is a cloud-based password manager developed by the same company behind NordVPN. It uses end-to-end encryption and supports syncing across multiple devices, though the free plan is limited to a single device.

For users already within the Nord ecosystem, NordPass can offer convenience and bundled pricing alongside VPN services. While not open source, it benefits from the security infrastructure and branding of an established privacy-focused provider.

NordPass is best suited for users seeking a simplified, bundled security solution rather than maximum configurability.

PROTON PASS

Proton Pass is an open-source password manager developed by Proton, known for its encrypted email and VPN services. It uses end-to-end encryption and has undergone independent security audits. Proton Pass integrates closely with the Proton ecosystem, making it attractive for users already using ProtonMail or ProtonVPN.

A free tier is available, with paid plans offering expanded features. Proton Pass is particularly appealing to users who want a unified, privacy-first ecosystem managed by a single provider.

CHOOSING THE RIGHT PASSWORD VAULT

There is no universally "best" password manager—only the best fit for your priorities. If open-source transparency and flexibility matter most, Bitwarden or KeePass are strong options.

If you want offline, self-managed security, KeePass offers maximum control. If you value polished usability, travel features, and managed security, 1Password is well suited.

If you prefer ecosystem integration, Proton Pass or NordPass may be convenient choices. Regardless of the tool you choose, a password vault should be viewed as one layer in a broader defense-in-depth strategy.

FINAL THOUGHTS: WHY PASSWORD VAULTS ARE ESSENTIAL FOR CRYPTO USERS

For cryptocurrency holders, password vaults significantly reduce risk by isolating credentials, preventing reuse, and encrypting sensitive data. They help protect exchange accounts from credential-stuffing attacks, secure wallet passphrases and recovery information, and limit the damage caused by third-party data breaches.

Strong vault usage should be paired with good hygiene. Use a long passphrase—not a simple password—as your master key. Enable two-factor authentication wherever possible. Avoid writing passwords down or storing them in unsecured locations.

By incorporating a secure password vault into your security strategy, you greatly reduce the likelihood of credential theft and strengthen your overall protection—keeping both personal identity and digital assets under your control.

TWO-FACTOR AUTHENTICATION (2FA): STRENGTHENING ACCOUNT SECURITY

The final critical component of personal device and account security is two-factor authentication (2FA). When used correctly, 2FA significantly reduces the risk of unauthorized access by adding an additional verification step beyond a password alone.

Authentication systems are built around three fundamental security factors:

- Something you know, such as a password, PIN, or secret answer
- Something you have, such as a mobile device, authenticator app, or physical security key
- Something you are, which includes biometric data like fingerprints, facial recognition, or retinal scans

Two-factor authentication requires any two of these factors to gain access to an account. In some environments, all three are used together—often referred to as three-factor authentication (3FA)—to provide even stronger protection.

HOW 2FA WORKS IN PRACTICE

To understand how these layers work together, consider a real-world example of three-factor authentication:

First, you unlock your computer using biometric authentication, such as a fingerprint or facial recognition—something you are. Next, you log into a website by entering your password—something you know. Finally, you confirm the login by entering a one-time code generated on your mobile device or security key—something you have.

Each step reinforces the others. Even if an attacker manages to steal your password, they still cannot access your account without possession of the second (or third) factor. This layered approach dramatically reduces the likelihood of successful account compromise.

WHY YOU SHOULD USE 2FA FOR CRYPTO & FINANCIAL ACCOUNTS

Cybercriminals consistently target financial accounts, and cryptocurrency holdings are among the most attractive assets to steal. As digital assets have increased in value, attackers have shifted their focus toward exploiting weak authentication practices—particularly accounts protected by passwords alone.

In response, many banks, email providers, and cryptocurrency exchanges now require or strongly encourage two-factor authentication as a baseline security measure. By requiring something you have in addition to something you know, 2FA adds a powerful barrier against unauthorized access—even when passwords are leaked, reused, or phished.

However, not all forms of 2FA provide the same level of protection. One of the most common weaknesses is SMS-based authentication. SIM swap attacks—where criminals hijack a victim's phone number to intercept verification codes—have become increasingly common, especially in high-value financial contexts. Once a phone number is compromised, attackers can bypass text-message-based 2FA entirely.

For cryptocurrency and financial accounts, best practices are clear:

- Avoid SMS-based 2FA whenever possible
- Use an authenticator app that generates time-based one-time codes
- For maximum security, use a physical hardware security key

Strong authentication is one of the most effective defenses against account takeover. In a decentralized financial environment where transactions are irreversible and responsibility rests entirely with the user, relying on passwords alone is no longer sufficient. Two-factor authentication is not an optional enhancement—it is a foundational requirement for protecting digital assets.

SOME TWO-FACTOR AUTHENTICATION APPS AND DEVICES WELL SUITED FOR SECURITY AND PRIVACY

Two-factor authentication significantly strengthens account security by requiring an additional verification step beyond a password. Authenticator apps generate time-based one-time passwords (TOTP) that change every few seconds, making them far more resistant to interception and replay attacks than SMS-based codes. While no solution is perfect, these tools represent a meaningful improvement over single-factor authentication when used correctly.

The following authenticator solutions are presented for educational purposes only. Each has advantages and limitations that should be evaluated in the context of personal risk tolerance, recovery planning, and overall security strategy.

Note: *None of the products discussed in this book are endorsements or recommendations. They are presented solely for educational purposes to illustrate various tools, concepts, and approaches to network and cybersecurity. Readers are encouraged to conduct their own research and due diligence to determine which products or solutions best meet their individual needs.*

Google Authenticator generates time-based one-time passwords compatible with a wide range of online services. Its design emphasizes simplicity and offline functionality, which reduces exposure to cloud-based attacks.

However, Google Authenticator does not provide built-in cloud backups. If a device is lost or replaced without transferring codes beforehand, account recovery can be difficult or impossible. Users relying on this app must plan migrations carefully and maintain alternative recovery options where available.

Best suited for: Users who want a lightweight, offline authenticator and are comfortable managing manual transfers and recovery planning.

Microsoft Authenticator supports both Microsoft and third-party accounts and includes optional cloud syncing, making account recovery easier when switching devices. It also offers additional features such as password autofill and account management tools.

Because it integrates with a broader account ecosystem, users should ensure that the associated Microsoft account is itself protected with strong authentication and recovery safeguards.

Best suited for: Users operating within the Microsoft ecosystem or those who prioritize easier device recovery and account continuity.

Authy provides encrypted cloud backups and supports synchronization across multiple devices, including desktop applications. These features can reduce the risk of permanent lockout due to device loss.

At the same time, cloud-backed recovery introduces additional dependency on account security. Protecting the Authy account with a strong password and additional safeguards is essential to prevent unauthorized access.

Best suited for: Users who value recovery flexibility and multi-device access while maintaining strong account protections.

LastPass Authenticator combines authentication functionality with password management features, offering a unified approach to securing logins. It supports encrypted storage, push notifications, and cross-device synchronization.

Because authentication is closely tied to the broader LastPass ecosystem, users should carefully evaluate their comfort with platform consolidation and ensure that the master account is exceptionally well secured.

Best suited for: Users already using LastPass who prefer consolidated security tools under a single platform.

PHYSICAL 2FA KEYS: THE MOST SECURE AUTHENTICATION OPTION

For users seeking the highest level of account security, physical hardware authentication keys represent the strongest form of two-factor authentication currently available. Unlike passwords, SMS codes, or app-generated one-time codes, hardware keys require **physical possession and direct user interaction** to approve a login.

These devices work by cryptographically proving possession of the key at the moment of authentication. Because the key must be physically present and touched or connected to the device, it cannot be phished through fake emails, duplicated by malicious websites, intercepted over a network, or bypassed through SIM swap attacks. Even if an attacker knows your password, access is impossible without the hardware key itself.

For cryptocurrency holders and anyone securing high-value financial accounts, this distinction is critical. Hardware keys remove entire categories of remote attacks by design, shifting security away from vulnerable networks and back into the user's direct control. In environments where transactions are irreversible and account recovery options are limited, this level of protection is difficult to overstate.

<u>**Note:**</u> *None of the products discussed in this book are endorsements or recommendations. They are presented solely for educational purposes to illustrate various tools, concepts, and approaches to network and cybersecurity. Readers are encouraged to conduct their own research and due diligence to determine which products or solutions best meet their individual needs.*

YUBIKEY – A WIDELY USED HARDWARE AUTHENTICATION DEVICE

YubiKey is one example of a hardware security key commonly used in professional and high-security environments. It supports multiple authentication standards and requires deliberate physical interaction—such as pressing a button—to complete authentication.

YubiKey devices are available in various forms and may support USB, NFC, or Bluetooth connections depending on the model. Because authentication occurs locally on the device, credentials are never transmitted in a way that can be reused or intercepted remotely.

Hardware keys like this are often used to secure email accounts, password vaults, cryptocurrency exchanges, and developer platforms where account compromise would carry significant consequences.

OTHER HARDWARE-BASED AUTHENTICATION OPTIONS

In addition to YubiKey, several other hardware-based authentication devices and standards exist, including:

- CryptoTrust keys
- Thetis security keys
- GoTrust devices
- FIDO U2F and FIDO2-compliant hardware keys

While implementations vary, the underlying security model remains the same: authentication requires physical possession of a trusted device.

BEST PRACTICE FOR HIGH-VALUE ACCOUNTS

For users managing significant cryptocurrency holdings or critical financial infrastructure, combining a hardware security key with strong passwords and authenticator apps provides the most resilient defense against account takeover. Hardware keys are not a replacement for good security hygiene, but they are an exceptionally strong final barrier when other protections fail.

In a decentralized financial system where responsibility rests entirely with the user, physical authentication keys represent a practical way to enforce true control—ensuring that access cannot occur without your direct, physical involvement.

CHAPTER 6 - SECURING YOUR NETWORK

NETWORK SECURITY: PROTECTING YOUR INTERNET CONNECTION

High-speed internet access has become nearly universal, with modern home networks often matching—or even exceeding—the speeds of business networks from just a decade ago. A single internet connection now supports streaming media, remote work, cloud services, smart home devices, voice-over-IP calls, and financial activity, all operating simultaneously within the same digital environment.

While home networking may appear simple on the surface, it represents one of the most critical—and frequently overlooked—layers of personal security. Every device connected to your network shares a common pathway to the internet, and weaknesses at the network level can expose otherwise well-secured computers, phones, and accounts.

For cryptocurrency users, this risk is especially significant. Network-level attacks can enable traffic interception, credential theft, malicious redirection, and unauthorized access to connected devices. Even strong passwords, encrypted wallets, and hardware authentication can be undermined if the underlying network is compromised.

Securing your home network is therefore not about paranoia—it is about containment and control. A properly configured network reduces the attack surface available to outsiders and limits the damage that can occur if one device becomes compromised. In a decentralized financial environment where users are fully responsible for their own security, network protection is a foundational layer that supports everything built on top of it.

In the sections that follow, we will examine common network vulnerabilities and practical steps you can take to harden your internet connection—helping ensure that your devices, data, and digital assets remain protected at the point where they are most exposed.

TYPES OF NETWORK CONNECTIONS: WIRED VS. WIRELESS

Home and small-office networks generally rely on two primary types of connections: wired (Ethernet) and wireless (Wi-Fi). Each plays a different role in network performance and security, and understanding their strengths and limitations is essential when protecting sensitive data and cryptocurrency activity.

WIRED (ETHERNET) CONNECTIONS

Wired connections use physical cabling—typically copper Ethernet cables or fiber-optic lines—to transmit data directly between devices. Most modern wired networks rely on RJ-45 Ethernet cables, which provide a stable, high-speed connection with minimal interference.

From a security perspective, wired connections offer several advantages. Because data travels through a physical cable rather than over the air, it is significantly harder for unauthorized parties to intercept traffic. Wired connections also tend to deliver lower latency and more consistent speeds, making them ideal for tasks that require reliability and security.

A typical wired setup works as follows: a coaxial or fiber line from an internet service provider connects to a modem. The modem then connects via Ethernet to a router. From the router, Ethernet cables distribute the connection to individual devices, and in many cases, the router also provides wireless access.

For cryptocurrency users, wired connections are generally the safest option for accessing exchanges, managing wallets, or performing sensitive financial operations.

WIRELESS (WI-FI) CONNECTIONS

Wireless networks transmit data using radio frequencies—most commonly in the 2.4 GHz and 5 GHz ranges. Wi-Fi provides flexibility and mobility, allowing laptops, phones, tablets, and smart devices to connect without physical cables.

While convenient, wireless connections introduce additional security considerations. Because Wi-Fi signals are broadcast through the air, they can potentially be intercepted by nearby attackers if the network is poorly configured. Weak passwords, outdated encryption standards, or misconfigured routers can expose wireless networks to unauthorized access.

In a typical wireless setup, the modem connects to a Wi-Fi router, which broadcasts a wireless signal that devices use to connect to the internet. Some modern devices combine the modem and router into a single unit, offering built-in Wi-Fi functionality.

Wireless networks can be made secure, but they require proper encryption, strong credentials, and ongoing configuration. For high-risk activities—such as managing cryptocurrency holdings—extra precautions should be taken when using Wi-Fi, especially on shared or public networks.

ROUTER & MODEM SECURITY: THE FIRST LINE OF DEFENSE

Your router and modem form the gateway between your private network and the public internet. Every device in your home—computers, phones, tablets, smart devices, and hardware wallets—relies on these components to communicate with the outside world. If this gateway is poorly secured, even well-protected devices can be exposed.

Because routers and modems operate continuously and often go years without attention, they are frequently targeted by attackers looking for an easy entry point into a network.

Change Default Credentials Immediately

Most routers and modem/router combo devices ship with default usernames and passwords. These credentials are widely documented online and are often the first thing attackers try when probing a network.

Leaving default credentials unchanged is equivalent to leaving your front door unlocked. Changing the administrative username and password as soon as the device is installed is one of the most important steps you can take to prevent unauthorized access.

Avoid ISP-Provided Routers When Possible

Internet service providers often supply basic modem/router combination devices for convenience. While functional, these devices typically offer limited security controls, infrequent firmware updates, and minimal customization options.

Using your own dedicated router—separate from the modem—provides greater control over security settings, stronger firewall options, and more frequent updates. This allows you to tailor your network configuration to your specific security needs rather than relying on a one-size-fits-all solution.

DISABLE UNUSED WIRELESS AND NETWORK FEATURES

Some routers, particularly those provided by ISPs, include features such as public Wi-Fi hotspots or default guest networks. While convenient for general use, these features can unintentionally expand your network's attack surface.

If you do not explicitly need a guest network or public hotspot feature, it should be disabled. Every unnecessary service running on your router represents another potential point of entry for attackers.

SECURE PHYSICAL ACCESS TO NETWORK EQUIPMENT

Network security is not purely digital. Physical access to a router or modem can allow an attacker to bypass software protections entirely by plugging directly into the device or resetting it to factory settings.

Routers and modems should be placed in a secure location where unauthorized individuals cannot access them. If someone can physically access your network hardware, many other security measures become irrelevant.

CONFIGURING YOUR NETWORK FOR MAXIMUM SECURITY

ENABLING AND USING A FIREWALL

A firewall acts as a gatekeeper for network traffic, blocking unauthorized inbound connections while allowing legitimate outbound communication. Most modern routers include a built-in firewall, but it is not always enabled by default.

Ensuring that your router's firewall is active provides a strong baseline level of protection against external scanning, intrusion attempts, and automated attacks. For users with more advanced needs, such as hosting services or managing multiple network segments, a dedicated hardware firewall may offer additional control and visibility.

It is important to note that firewall configurations may require customization if you host personal servers, use static IP services, or rely on specialized network applications. In such cases, rules should be adjusted carefully to allow necessary traffic without exposing the broader network.

KEY TAKEAWAY

Your router and modem represent the first and most critical security boundary between your private devices and the internet. Proper configuration—strong credentials, minimal exposed features, physical security, and an active firewall—dramatically reduces the likelihood of unauthorized access.

In a layered security model, protecting the network gateway strengthens every device and account behind it, providing a solid foundation for safeguarding personal data and cryptocurrency assets alike.

UNDERSTANDING WI-FI SECURITY & ENCRYPTION STANDARDS

Wireless networking is governed by technical standards developed by the Institute of Electrical and Electronics Engineers (IEEE). These standards, collectively known as the 802.11 family, define how Wi-Fi networks transmit data, which frequency bands they use, and their maximum performance capabilities.

While speed and coverage often receive the most attention, the security implications of these standards are just as important—particularly when wireless networks are used to access financial accounts or manage cryptocurrency assets.

WI-FI NETWORK STANDARDS (802.11)

Over time, Wi-Fi technology has evolved significantly. Older standards remain supported for compatibility, but they offer limited performance and, in some cases, weaker security assumptions.

- 802.11a operates on the 5 GHz band and supports speeds up to 54 Mbps. It is largely obsolete and rarely used today.
- 802.11b runs on the 2.4 GHz band with a maximum speed of 11 Mbps and should be considered outdated.
- 802.11g improved performance on the 2.4 GHz band to 54 Mbps but is no longer suitable for modern networks.
- 802.11n introduced dual-band support (2.4 GHz and 5 GHz) with speeds up to 600 Mbps, making it common in older home networks.
- 802.11ac operates on the 5 GHz band and supports speeds exceeding 1 Gbps, making it a strong choice for modern home and business environments.
- 802.11ax, also known as Wi-Fi 6, supports both 2.4 GHz and 5 GHz bands and significantly improves speed, efficiency, and device handling in crowded networks.

For most users today, 802.11ac remains a solid and widely supported standard. Wi-Fi 6 (802.11ax) represents the current generation and offers improved performance, especially in environments with many connected devices.

SECURING YOUR WI-FI NETWORK: ENCRYPTION MATTERS

Wi-Fi encryption determines how data is protected as it travels between devices and the router. Without proper encryption, nearby attackers can intercept traffic, capture credentials, or gain unauthorized access to the network.

Over the years, several Wi-Fi encryption protocols have been introduced:

- WEP (Wired Equivalent Privacy) is fundamentally broken and can be cracked in minutes using widely available tools. It should never be used.

- WPA (Wi-Fi Protected Access) was an interim improvement over WEP but is now considered obsolete.
- WPA2 introduced strong encryption and remains widely used and secure when properly configured.
- WPA3 further improves security by protecting against brute-force attacks and improving encryption for modern devices. It represents the current best practicc where supported.

If your router still offers WEP or WPA as options, they should be avoided entirely. WPA2 or WPA3 should always be used for any network carrying sensitive data.

BEST PRACTICES FOR WI-FI SECURITY

Even with modern encryption, Wi-Fi networks require proper configuration to remain secure.

Use WPA2 or WPA3 encryption and ensure the password is long, unique, and difficult to guess. Change the default network name (SSID) to something that does not reveal personal information, addresses, or router models. Features such as Wi-Fi Protected Setup (WPS) should be disabled, as they are frequently exploited to bypass strong passwords.

Broadcasting your SSID is generally acceptable for home use, but if you choose to disable it, understand that this offers only minimal security benefit and should not be relied upon as a primary defense. Encryption and strong credentials remain far more important.

KEY TAKEAWAY

Wireless networking offers convenience and mobility, but it also introduces additional exposure compared to wired connections. Using modern Wi-Fi standards, strong encryption, and careful configuration significantly reduces the risk of unauthorized access.

For cryptocurrency users, Wi-Fi security is not just about faster speeds or better coverage—it is about ensuring that sensitive activity is not exposed over a poorly protected wireless connection. In a layered security model, properly secured Wi-Fi serves as a critical barrier between private assets and public airwaves.

FINAL THOUGHTS: STRENGTHENING YOUR NETWORK SECURITY

Your home network is the foundation upon which every connected device relies. Securing it properly reduces the risk of unauthorized access, data theft, and network intrusion—protecting not only everyday internet activity, but also sensitive financial and cryptocurrency operations.

At a minimum, a secure network should include the following core protections:

- Change default router and Wi-Fi administrative credentials immediately
- Enable modern Wi-Fi encryption (WPA2 or WPA3) and disable outdated protocols such as WEP and WPA
- Ensure the router's firewall is enabled
- Physically secure network hardware to prevent tampering or unauthorized connections
- Consider upgrading to newer Wi-Fi standards, such as Wi-Fi 6, for improved performance and security

When both wired and wireless connections are properly secured, the entire network becomes far more resistant to external threats.

SECURING YOUR WI-FI NETWORK: CONFIGURATIONS & BEST PRACTICES

Understanding Wi-Fi standards and encryption methods is only the first step. Effective security depends on how those tools are configured in practice.

Regardless of whether a network uses older standards like 802.11n or newer ones such as 802.11ac or 802.11ax, encryption is what truly protects wireless traffic. Using WPA2 with AES encryption—or WPA3 where available—is essential. Without strong encryption, even a fast and modern network remains vulnerable.

That said, encryption alone is not enough. A weak Wi-Fi password undermines even the strongest encryption. If a password is easily guessed, attackers do not need to break encryption at all—they simply log in.

CREATING A STRONG WI-FI PASSWORD

Your Wi-Fi password is the first line of defense against unauthorized access. A strong password should:

- Be at least 12 characters long (longer is better)
- Avoid common dictionary words or predictable patterns
- Use a mix of uppercase and lowercase letters, numbers, and symbols
- Never rely on obvious choices such as names, addresses, or simple sequences

Weak passwords negate the benefits of strong encryption and leave the network exposed.

USING A PASSPHRASE: AN EASY WAY TO CREATE A STRONG PASSWORD

One effective way to create a strong but memorable Wi-Fi password is by using a passphrase. A passphrase starts with a sentence or quote and is transformed into a complex password through deliberate modification.For example, let's take a famous movie quote:

"Frankly, my dear, I don't give a damn." (*Gone with the Wind*)

We can modify it into a strong password:

1. Take the first letter of each word → Fmdidgad
2. Add an exclamation point → Fmdidgad!
3. Insert a symbol to break it up → Fmd;idgad!
4. Add a number for extra security → Fmd;1dgad!
5. Replace a letter with another symbol for added complexity → Fmd;1dg@d!

Final Password: Fmd;1dg@d! (10 characters, uppercase & lowercase letters, numbers, and symbols)

WHY THIS WORKS:

✅ Memorable: You can recall it by remembering the phrase.

✅ Difficult to guess: Even if someone knows the phrase, they won't easily guess your modifications.

✅ Meets security standards: Contains letters, numbers, and symbols in a randomized format.

FINAL THOUGHTS: WHY A STRONG WI-FI PASSWORD MATTERS

A properly secured Wi-Fi network ensures that:

- Your home network remains private and inaccessible to outsiders
- Cryptocurrency transactions and financial activity are protected in transit

- Unauthorized users cannot exploit connected devices as entry points

When strong encryption is combined with a robust passphrase, your wireless network becomes a hardened environment rather than an open doorway. In a decentralized financial world—where responsibility rests entirely with the user—network security is not optional. It is a fundamental requirement for protecting digital assets and maintaining long-term control.

CHAPTER 7 - FIREWALLS AND VPN'S

FIREWALLS & VPNS: STRENGTHENING NETWORK SECURITY

As your own bank and financial custodian, protecting your network infrastructure is a critical responsibility. While you may have limited control over the modem or gateway provided by your internet service provider, you retain full control over everything behind it—your router configuration, connected devices, wireless network, and the traffic flowing through them.

Firewalls and Virtual Private Networks (VPNs) serve as powerful tools in this layer of defense. They help restrict unauthorized access, limit exposure to external threats, and reduce the risk of interception or manipulation of sensitive data. When properly configured, these technologies form an essential barrier between your private environment and the public internet.

FIREWALLS: YOUR NETWORK'S FIRST LINE OF DEFENSE

A firewall is a security system that monitors and filters incoming and outgoing network traffic according to predefined rules. Its purpose is simple but essential: allow legitimate communication while blocking unauthorized or suspicious activity.

Without a firewall, devices on a network are far more exposed to scanning, probing, and intrusion attempts originating from the internet. With a firewall in place, only traffic that meets explicitly defined criteria is allowed through, significantly reducing the attack surface available to outsiders.

One way to think about a firewall is as controlled access rather than open exposure. Instead of allowing any external connection to reach your devices, a firewall enforces boundaries—permitting only what is necessary and denying everything else by default.

WHY FIREWALLS MATTER FOR CRYPTO USERS

For individuals managing cryptocurrency assets, the stakes are higher than average. Network-level compromise can enable attackers to monitor traffic, exploit vulnerable services, or gain footholds that bypass application-level protections.

A properly configured firewall helps:

- Block unsolicited inbound traffic from the internet
- Prevent automated scanning and intrusion attempts
- Limit exposure of internal devices and services
- Contain damage if a single device becomes compromised

In a decentralized financial environment where transactions are irreversible, preventing access is far more effective than attempting recovery after a breach.

IMPLEMENTING FIREWALL PROTECTION

Most consumer routers include a basic firewall that can be enabled through the device's settings. Ensuring this feature is active provides a baseline level of protection and should be considered mandatory rather than optional.

For users with greater security needs, a dedicated firewall appliance can offer enhanced visibility, finer-grained control, and more advanced filtering capabilities. These devices allow administrators to define strict rules around inbound and outbound traffic, creating a more hardened network perimeter.

Firewall configurations should be designed with a "deny by default" mindset—allowing only the traffic that is explicitly required for normal operation. Any unnecessary open ports or services increase exposure and should be avoided.

IS A DEDICATED FIREWALL EXCESSIVE?

For casual internet use, a dedicated firewall may be unnecessary. However, for individuals protecting substantial financial assets—including significant cryptocurrency holdings—the cost-benefit analysis changes.

Investing a few hundred dollars in dedicated network security hardware is a reasonable and proportionate measure when weighed against the potential impact of a successful breach. Security should scale with value, and a phased approach—adding protective layers as assets grow—is a practical and responsible strategy.

KEY TAKEAWAY

Firewalls establish the first meaningful boundary between your private network and the internet. Whether implemented through a router's built-in features or a dedicated device, they significantly reduce exposure to external threats and form a foundational layer of network defense.

In a layered security model, firewalls do not stand alone—but without them, every other protection becomes easier to bypass. Securing the network perimeter is not about paranoia; it is about control, containment, and long-term resilience.

WHAT IS A VPN? (VIRTUAL PRIVATE NETWORK)

A Virtual Private Network (VPN) is a technology designed to enhance privacy and security by encrypting your internet traffic and routing it through an intermediary server. When properly used, a VPN makes your online activity significantly more difficult to monitor or intercept by third parties.

Without a VPN, your internet traffic is visible at multiple points along its path. Internet service providers can see which websites you visit, network operators can inspect unencrypted traffic, and attackers on public Wi-Fi networks may attempt to intercept data. A VPN reduces this exposure by encrypting traffic before it leaves your device.

VPNs are commonly used to protect against:

- Internet service provider (ISP) tracking
- Traffic interception on public or unsecured Wi-Fi networks
- Broad forms of network-level monitoring and surveillance

HOW VPNS WORK

A VPN functions by creating an encrypted tunnel between your device and a VPN server operated by the provider. Once connected, all internet traffic is routed through this tunnel before reaching its final destination.

In practical terms, this involves three components:

- Your device (computer, phone, or tablet)
- A VPN server (often located in another city or country)
- The destination website or service (such as an exchange, email provider, or financial platform)

Without a VPN, your internet service provider can observe your browsing activity, websites can see your real IP address and approximate location, and attackers on shared networks may be able to intercept traffic.

With a VPN enabled:

- Your data is encrypted before leaving your device
- Your real IP address is hidden from destination websites
- Your connection remains protected even on unsecured networks

While a VPN does not make you anonymous, it significantly reduces passive monitoring and interception risks.

TYPES OF VPNS

VPNs are generally divided into two categories based on their original use cases and design goals.

Private Business VPNs (Corporate Use)

VPN technology was originally developed for enterprise environments. Business VPNs allow organizations to securely connect remote offices or employees over public networks while maintaining access to internal systems.

In this model, the VPN creates a secure tunnel between company locations or between employees and a centralized corporate network. These VPNs prioritize controlled access, authentication, and internal network security rather than anonymity.

CONSUMER VPNS (PRIVACY & SECURITY FOR INDIVIDUALS)

In recent years, consumer VPN services have become widely used by individuals seeking greater privacy and protection online. These services are designed to encrypt personal internet traffic and mask IP addresses rather than provide access to private corporate networks.

Consumer VPNs are commonly used for:

- Reducing online tracking and data collection
- Protecting financial and cryptocurrency activity on public networks
- Adding a layer of privacy when browsing
- Accessing region-restricted content

For cryptocurrency users, consumer VPNs are often used to reduce exposure when accessing exchanges, wallets, or financial services—particularly on shared or mobile networks.

HOW CONSUMER VPNS WORK

When you subscribe to a consumer VPN service, the process typically works as follows:

1. You install a VPN application on your device.
2. You select a VPN server location.
3. Your device establishes an encrypted tunnel to that server.
4. All internet traffic passes through the VPN server before reaching the public internet.

To outside observers, your traffic appears to originate from the VPN server rather than your real location.

For example, if multiple users connect to the same VPN server while visiting the same website, each user's traffic is encrypted and routed through the VPN's exit point. Although the users have different real IP addresses, the destination website only sees the VPN server's address. This makes it difficult to distinguish individual users based solely on network information.

KEY TAKEAWAY

A VPN encrypts your internet traffic and obscures your real IP address, reducing exposure to interception, tracking, and network-level monitoring. While it is not a substitute for strong authentication, firewalls, or secure devices, a VPN provides an important additional layer of protection—especially when accessing financial or cryptocurrency services on untrusted networks.

In a layered security model, VPNs help protect data in transit, complementing firewalls that protect network boundaries and device-level security controls that protect endpoints.

WHY USE A VPN FOR CRYPTO SECURITY?

Cryptocurrency activity often involves accessing exchanges, wallets, and financial services over the internet—sometimes from shared or mobile networks. In these situations, a VPN can provide meaningful protection by reducing exposure at the network level.

Using a VPN helps by encrypting internet traffic, which prevents internet service providers from passively tracking browsing behavior and reduces the risk of data interception on public or unsecured Wi-Fi networks. By masking your real IP address, a VPN also makes it more difficult for attackers to profile, target, or geographically identify cryptocurrency users based on network metadata alone.

For crypto holders, this added layer of privacy can be especially valuable when managing assets outside a trusted home network. However, it is important to understand the limits of VPNs. While a VPN protects data in transit, it does not replace wallet-level encryption, secure authentication, or proper key management. Your wallet security remains the most critical safeguard.

A VPN should therefore be viewed as a complementary layer—not a standalone solution.

FINAL THOUGHTS: STRENGTHENING YOUR DIGITAL SECURITY

Strong digital security is built through layered defenses rather than reliance on a single tool. Firewalls and VPNs serve different but complementary roles in this model.

A firewall controls which traffic is allowed to enter or leave your network, acting as a gatekeeper at the perimeter. A VPN encrypts traffic as it travels across the internet, protecting data from interception and reducing visibility into your online activity.

In practice, this layered approach can scale with your needs:

- A basic setup may involve enabling the built-in firewall on your router.
- A more advanced configuration could include a dedicated firewall appliance for tighter control.
- A VPN adds an additional privacy layer by encrypting traffic and masking your IP address during sensitive activity.

The objective is not invisibility, but reduced exposure—limiting opportunities for attackers and protecting financial activity from unnecessary scrutiny.

UNDERSTANDING HOW VPNS KEEP YOUR TRAFFIC SECURE

To better understand how VPNs protect users, it helps to look at how traffic is managed behind the scenes.

When multiple users connect to the same VPN server, all of their traffic exits the internet through a single public IP address. To ensure that responses are sent back to the correct user, the VPN server maintains a translation table that tracks each connection independently.

This process is similar to network address translation (NAT). Each client is assigned a unique port number on the VPN server, allowing the server to distinguish between multiple encrypted connections using the same public IP address.

For example, consider three users connecting to the same VPN server:

- One client may be assigned port 500
- Another port 800
- A third port 9000

Although all traffic appears to originate from the same VPN server IP to the outside world, the VPN server uses its internal mapping table to route return traffic back to the correct encrypted tunnel and device.

This mechanism ensures that:

- Each user's traffic remains logically separated
- Responses are delivered to the correct client
- Individual identities remain obscured from destination websites

KEY TAKEAWAY

VPNs protect data in transit by encrypting traffic and obscuring real IP addresses, making interception and passive monitoring far more difficult. When combined with firewalls, secure devices, strong authentication, and proper wallet management, VPNs strengthen the overall security posture of cryptocurrency users.

In a decentralized financial environment where responsibility rests with the individual, reducing exposure at every layer—device, network, and transit—is essential for long-term asset protection.

ROUTING TRAFFIC THROUGH THE VPN

Once a VPN has established encrypted tunnels and created its internal translation table, it can begin routing client traffic to destination websites. At this point, all data leaving the user's device is encrypted and encapsulated inside the VPN tunnel.

When a client accesses a cryptocurrency exchange, the process works as follows:

Each client sends an encrypted request from their device to the VPN server. Although the users have different real IP addresses, their traffic is tunneled to the same VPN endpoint. The VPN server then forwards the requests to the destination website using its own public IP address and a unique port assignment for each client.

From the perspective of the crypto exchange, all incoming traffic appears to originate from the VPN server—not from the individual users behind it. The exchange cannot see the real client IP addresses, locations, or network details.

This mechanism allows multiple users to share a single VPN exit point while keeping their sessions logically separated and protected.

RETURNING DATA TO CLIENTS

When the crypto exchange sends a response, it replies to the VPN server's public IP address along with the specific port number associated with each session. The VPN server uses its translation table to determine which encrypted tunnel the response belongs to.

The VPN server then re-encrypts the response and sends it back through the correct tunnel to the originating client device. Each user receives only their own data, even though all traffic flows through the same VPN server.

Throughout this process:

- The destination website never sees the real IP addresses of individual users
- The VPN server manages session separation internally using port assignments
- Traffic remains encrypted between the client device and the VPN server

HOW A VPN PROTECTS YOU ON PUBLIC WI-FI

Public Wi-Fi networks—such as those found in coffee shops, airports, hotels, or libraries—are inherently insecure. These networks are shared by dozens or even hundreds of users, and in many cases, anyone connected can potentially monitor network traffic using readily available tools.

Without a VPN, data sent over public Wi-Fi may be visible to attackers running traffic sniffers or performing man-in-the-middle attacks. This can expose sensitive information such as login credentials, session cookies, and financial activity. For cryptocurrency users, this risk is especially serious, as stolen credentials can lead to irreversible loss.

When a VPN is enabled, all traffic is encrypted before it leaves your device. Even if an attacker captures the data packets traveling across the public network, the information remains unreadable. This effectively neutralizes many common public Wi-Fi attacks by ensuring that intercepted data cannot be interpreted or reused.

It is important, however, to understand the scope of this protection. A VPN encrypts traffic between your device and the VPN server. Once traffic exits the VPN server and travels to the final destination—such as a cryptocurrency exchange or banking website—it is only encrypted if the destination site uses HTTPS. Fortunately, reputable financial and crypto platforms do use HTTPS by default, but this distinction is still important to understand.

KEY TAKEAWAY

Public Wi-Fi should always be treated as hostile. A VPN does not make a network "safe," but it dramatically reduces the risk of data interception by encrypting traffic at its most vulnerable point—between your device and the internet.

For cryptocurrency users, enabling a VPN on public networks is not a convenience feature; it is a basic defensive measure that helps prevent credential theft, account compromise, and financial loss.

SOME VPN PROVIDERS WELL SUITED FOR PRIVACY AND SECURITY

As concerns around online privacy and data collection continue to grow, many VPN providers now offer more than basic traffic encryption. Some services include additional tools such as encrypted email, cloud storage, or password managers, creating broader privacy-focused ecosystems.

Selecting a VPN service depends on individual priorities, including trust model, jurisdiction, performance needs, and usability. The following providers are discussed for informational purposes only, not as endorsements, and are presented in no particular order.

Note: *None of the products discussed in this book are endorsements or recommendations. They are presented solely for educational purposes to illustrate various tools, concepts, and approaches to network and cybersecurity. Readers are encouraged to conduct their own research and due diligence to determine which products or solutions best meet their individual needs.*

PROTONVPN

ProtonVPN is developed by the same organization behind Proton's encrypted email and privacy tools. It operates under Swiss jurisdiction, which offers strong privacy protections and is outside many international data-sharing alliances.

ProtonVPN offers a free tier with limited speeds and server locations, as well as paid plans that include higher performance and additional features. The service maintains a stated no-logs policy and emphasizes transparency and user privacy.

Well suited for: Users seeking a privacy-focused VPN provider, including those who want to explore VPN use through a limited free option.

NORDVPN

NordVPN is a widely used consumer VPN known for offering a broad set of features. These include options such as multi-hop routing, obfuscated servers designed to bypass certain network restrictions, and specialized servers for different use cases.

The service provides applications across major platforms and is often chosen for its balance of performance and configuration flexibility. As with any VPN provider, users should evaluate policies, jurisdiction, and account security carefully.

Well suited for: Users who want a wide range of features and configurable security options with strong performance.

EXPRESSVPN

ExpressVPN emphasizes ease of use and performance, offering applications designed to work with minimal configuration. The provider uses a RAM-only server architecture, meaning data is not written to persistent storage on servers.

This approach reduces the risk of retained data but places trust in the provider's operational practices. ExpressVPN is often selected by users who want strong defaults without extensive configuration.

Well suited for: Users who prioritize simplicity, speed, and a streamlined VPN experience.

IPVANISH

IPVanish offers support for an unlimited number of devices per account and provides tools intended to work in restrictive network environments. It emphasizes encryption and user control across multiple platforms.

As with all VPN services, users should review logging policies and account security practices to ensure alignment with their privacy expectations.

Well suited for: Users managing many devices or operating in environments with restrictive network controls.

FINAL THOUGHTS: WHY VPNS ARE ESSENTIAL FOR CRYPTO SECURITY

VPNs play an important role in protecting cryptocurrency users by encrypting internet traffic and reducing exposure at the network level. When properly configured, a VPN can:

- Encrypt traffic to protect sensitive activity in transit
- Mask real IP addresses, reducing profiling and targeted attacks

- Secure connections on public or shared Wi-Fi networks
- Limit passive monitoring by internet service providers

That said, VPNs are not a replacement for secure wallets, strong authentication, or proper key management. They are most effective when integrated into a broader layered-security strategy.

For best results, VPN use should be combined with:

- Strong, unique passwords and multi-factor authentication
- A properly configured firewall
- Caution when using public networks
- Secure devices and updated software

When used appropriately, a VPN helps reduce exposure and improve privacy—but it is only one layer in a comprehensive approach to digital and cryptocurrency security.

TRANSITION: FROM DIGITAL ENVIRONMENT TO DIGITAL ASSETS

Up to this point, we have focused on securing the environment in which cryptocurrency is accessed and managed—your devices, browsers, networks, accounts, and online behavior. These layers form the foundation of crypto security. If the environment is compromised, even the strongest wallet or protocol-level protections can be rendered meaningless.

From here forward, the focus shifts. The next chapters move inside the crypto system itself: wallets, private keys, seed phrases, exchanges, and custody choices. This is where ownership is defined, where control is enforced, and where mistakes become irreversible. While computer security protects the *conditions* under which crypto is used, wallet and key management protect the *assets themselves*.

Together, these two halves form a complete security model. The first defends the perimeter. The second safeguards the vault. Understanding both—and how they work together—is essential to taking full responsibility for your role as your own bank.

CHAPTER 8 - A WALLET FOR YOUR CRYPTO

UNDERSTANDING CRYPTOCURRENCY WALLETS

Whether you realize it or not, if you are interacting with cryptocurrency, you are using a wallet. Wallets are not an optional feature of crypto—they are a foundational component of how digital assets are owned, transferred, and secured.

For users who operate exclusively through centralized exchanges (CEXs) such as Coinbase, this reality is often hidden. In these environments, the exchange manages wallets on your behalf, abstracting away the technical details. While this can make getting started easier, it can also create a false sense of ownership and security.

At its core, a cryptocurrency wallet is not a container that *holds* crypto or coins. Instead, it is a tool that manages cryptographic keys—keys that prove ownership and grant permission to move funds recorded on the blockchain. Understanding this distinction is critical, because in crypto, **control of the wallet means control of the assets**.

This chapter will explore what wallets actually are, how they function, and why choosing the right wallet—and securing it properly—is one of the most important decisions a crypto user can make. We will examine:

- **Types of wallets** — including hot wallets, cold wallets, hardware wallets, and software wallets
- **Wallet security fundamentals** — private keys, public keys, seed phrases, and passphrases
- **How wallets really work** — clearing up common misconceptions about where crypto "lives" and who truly controls it
- **How wallets interact with the Blockchain**

As cryptocurrency shifts responsibility away from institutions and toward individuals, wallets become the center of personal security. Understanding them is not just a technical exercise—it is the difference between custodial convenience and true financial sovereignty.

WHAT DOES A WALLET ACTUALLY HOLD?

One of the most common misconceptions about cryptocurrency wallets is the belief that they *hold* or *store* cryptocurrency in the same way a traditional wallet holds cash or cards. This analogy is understandable—but ultimately misleading.

In reality, a cryptocurrency wallet does not store coins or tokens at all. Instead, it stores **cryptographic keys** which provide a means to interact with the blockchain. These keys allow you to view balances, prove ownership, and authorize transactions. The wallet is best thought of as a **control interface**, not a container.

Your cryptocurrency never leaves the blockchain. What changes is *who has the ability to access and move it*.

Wallets are simply tools that allow you to interact with it. Understanding this distinction is essential. In cryptocurrency, **ownership is not determined by who you are, but by what keys you control**. Lose the keys, and you lose access. Control the keys, and you control the assets—regardless of where you are or who you are.

At a practical level, a wallet performs three essential functions:

- It manages your public and private keys
- It communicates with the blockchain to check balances and transaction history
- It allows you to authorize transactions to send and receive cryptocurrency

The wallet is therefore a control mechanism, not a storage container.

A cryptocurrency wallet holds **private keys**—and nothing else of value. These private keys are the cryptographic credentials that give you control over specific funds recorded on the blockchain.

Private keys serve several critical functions. They identify which blockchain addresses you control, they generate the cryptographic signatures required to authorize transactions, and they allow you to reassign ownership of cryptocurrency from one address to another. Without the private key, you cannot move funds—even though the blockchain may show that those funds exist.

This is why private keys are so powerful, and why protecting them is essential. Anyone who gains access to your private key gains the same control over your funds that you have.
This distinction becomes especially clear with hardware wallets. Even when a hardware wallet is completely disconnected from the internet—or powered off entirely—your cryptocurrency remains fully intact on the blockchain. Funds can still be sent to your wallet's address at any time, because the blockchain records transactions independently of your device.

For example, if someone sends Bitcoin to your hardware wallet's address while the device is unplugged, the transaction is still confirmed and recorded on the blockchain. Nothing is "stored" in the wallet itself. When you later reconnect the device and open the associated software, the wallet simply scans the blockchain, detects the transaction, and updates the displayed balance.

A USEFUL WAY TO THINK ABOUT WALLETS

Rather than imagining a wallet as a container, think of it as a **keychain**.

The blockchain is the vault where all cryptocurrency exists. Your wallet holds the keys that unlock your ability to interact with specific entries in that vault. Lose the keys, and access is gone—even though the funds remain recorded on the blockchain forever.

EXAMPLE: PAYING WITH BITCOIN

When you purchase something with Bitcoin, no coins are physically transferred from your wallet to the vendor. Instead, your wallet uses your private key to **authorize a transaction** that reassigns ownership on the blockchain.

In practical terms, this means:

- Your wallet signs a transaction using your private key
- The transaction specifies that a portion of Bitcoin associated with your address should now belong to the recipient's address
- The blockchain network verifies the signature and confirms the transaction
- The transaction is permanently recorded on the blockchain

A transaction fee is included as part of this process. This fee compensates miners (or validators, depending on the network) for verifying and processing the transaction.

KEY TAKEAWAY

A wallet does not hold cryptocurrency—it holds the **authority** to move it. Ownership in cryptocurrency is enforced by cryptography, not by possession, accounts, or institutions. Control the private key, and you control the funds. Lose it, and access is gone—regardless of what the blockchain ledger says.

This is why wallets sit at the center of crypto security—and why everything in this chapter builds toward understanding and protecting private keys.

In crypto, ownership is not based on possession of a device or an account—it is based entirely on control of cryptographic keys. Understanding this distinction is foundational to security, self-custody, and true financial sovereignty.

THE BLOCKCHAIN: WHERE YOUR CRYPTO ACTUALLY EXISTS

Unlike physical money, cryptocurrency exists only as records on a blockchain. The blockchain is a decentralized, public ledger that records every transaction ever made on that network. Each transaction updates balances associated with specific addresses, and those records are shared and verified across thousands of independent computers worldwide.

Because the blockchain is public, anyone can verify any transactions at any time. While wallet addresses are pseudonymous—meaning they are not directly tied to real-world names or identities—the transaction data itself is fully transparent.

To illustrate this, consider the very first Bitcoin transaction. The address that received the original block reward in January 2009 is still visible today. Anyone can look it up on a public blockchain explorer and see that it received 50 newly generated bitcoins at the network's launch. Although the identity of the owner is unknown, the transaction itself is permanently recorded and verifiable.

This transparency is a defining feature of blockchain systems. Ownership is not established through names, accounts, or institutions—it is established through control of cryptographic keys that correspond to blockchain addresses.
To understand how cryptocurrency wallets function, it helps to look at the very beginning of Bitcoin itself.

On January 3, 2009, the first Bitcoin block—known as the Genesis Block—was created. As part of that block, 50 BTC were issued as "newly generated coins" and assigned to a specific wallet address. At the time, those coins had no monetary value. Today, that same 50 BTC would be worth millions of dollars.

What matters is not the value, but how that transaction exists. That original transaction is still recorded on the Bitcoin blockchain today. It has never moved, never disappeared, and never needed to be "stored" anywhere else. The blockchain simply records that a specific address was credited with 50 BTC at that moment in time. Ownership has always been determined by control of the cryptographic keys associated with that address—not by possession of a wallet device or application.

This is how all cryptocurrency wallets work. When you receive crypto, the blockchain records that a certain amount is associated with your wallet's address. When you send crypto, your wallet uses your private key to authorize a transaction that updates the blockchain's ledger. The wallet does not move coins—it signs messages that tell the network how ownership should change.

Understanding this is essential: wallets manage permissions, blockchains record reality.

HOW TO VIEW A TRANSACTION ON THE BITCOIN BLOCKCHAIN

1. Go to: Blockchain Explorer (Https://www.blockchain.com/explorer)
2. Enter this wallet address: 1A1zP1eP5QGefi2DMPTfTL5SLmv7DivfNa
3. Click Search → Select BTC for Bitcoin results.

WHAT YOU'LL SEE:

This wallet received the first-ever Bitcoin transaction, containing 50 BTC as "Newly Generated Coins" on January 3, 2009.

KEY TAKEAWAY:

Every Bitcoin transaction ever made can be found on the blockchain. While you can't see the owner's identity, the transaction details are fully transparent and publicly accessible. The blockchain is the source of truth.

UNDERSTANDING PRIVATE KEYS & SEED PHRASES

The security of your cryptocurrency ultimately comes down to two critical elements: your private key and your seed phrase. Everything else—wallet software, hardware devices, exchanges, and apps—exists only to help manage and protect these two things.

A private key is a unique cryptographic string that proves ownership of your cryptocurrency. It allows you to authorize transactions and move funds on the blockchain. Whoever controls the private key controls the funds—no exceptions.

A seed phrase (sometimes called a recovery phrase or mnemonic phrase) is a human-readable backup of your wallet. It is typically a random list of 12 to 24 words generated when a wallet is first created. This phrase acts as a master key, from which all private keys for that wallet can be mathematically derived.

If your device is lost, damaged, or destroyed, the seed phrase is what allows you to restore access to your funds on a new device. If the seed phrase is lost, recovery is impossible.

WHY SEED PHRASES MATTER SO MUCH

A seed phrase is not just a backup—it **is the wallet**.

Anyone who gains access to your seed phrase can recreate your wallet, generate the same private keys, and transfer all associated funds elsewhere. There is no warning, no approval step, and no recovery process once this happens.

This leads to the most important rules in cryptocurrency security:

THE THREE RULES OF CRYPTO SECURITY

🚨 **Never share your seed phrase or private key with anyone.**

🚨 **Never share your seed phrase or private key with anyone.**

🚨 **Never share your seed phrase or private key with anyone.**

No legitimate wallet provider, exchange, support agent, or developer will ever need your seed phrase. Anyone asking for it—under any circumstance—is attempting to steal your funds.

BEST PRACTICES FOR PROTECTING SEED PHRASES

To reduce risk and ensure long-term security:

✅ Write your seed phrase down **by hand** and store it in a secure physical location

✅ Never store your seed phrase in screenshots, notes apps, cloud storage, or email

✅ Use a hardware wallet to isolate private keys from internet-connected devices

✅ Only enter your seed phrase when **restoring a legitimate wallet you control**

KEY TAKEAWAY

In traditional finance, access can be reset, frozen, or recovered. In cryptocurrency, **ownership is absolute**. The seed phrase is the ultimate authority over your funds, and protecting it is not optional—it is foundational.

If you understand nothing else about wallet security, understand this:

Lose your seed phrase, and you lose your crypto.
Share your seed phrase, and someone else owns it.

Everything else in crypto security builds on this principle.

PROTECTING YOUR CRYPTOCURRENCY WALLET: PRIVATE KEYS & SEED PHRASES

By now, it should be clear that nothing in cryptocurrency security is more important than protecting your private keys and seed phrase. These are not just technical details—they are the *sole proof of ownership* of your digital assets.

CRITICAL RULE:
ANYONE WHO HAS ACCESS TO YOUR WALLET'S SEED PHRASE OR PRIVATE KEY HAS COMPLETE CONTROL OVER YOUR CRYPTOCURRENCY.

There are no safeguards once funds are moved. Blockchain transactions are irreversible by design. If your crypto is transferred out of your wallet without your permission, there is no recovery process, no fraud department, and no customer support to undo the loss.

HOW CRYPTO THEFT HAPPENS

If a scammer, hacker, or malicious actor gains access to your private key or seed phrase, they can:

- Recreate your wallet on their own device
- Instantly authorize transactions as if they were you

- Transfer all of your funds to another wallet
- Permanently empty your holdings—often in seconds

Because blockchain networks do not rely on identity, ownership is determined purely by cryptographic proof. The network cannot tell the difference between you and an attacker using your keys.

HOW WALLET KEYS WORK TOGETHER

To understand why this is so dangerous, it helps to see how the components of a wallet interact:

- Private Key – The cryptographic secret that authorizes transactions and proves ownership
- Seed Phrase – A human-readable backup that can regenerate all private keys for the wallet
- Public Key / Address – The shareable identifier that allows others to send you funds

THE BOTTOM LINE

In traditional finance, access can be frozen, reversed, or restored. In cryptocurrency, **ownership is absolute**.

Everything else in crypto security—hardware wallets, 2FA, firewalls, and VPNs—exists to protect this single point of failure. Treat your private keys and seed phrase with the same seriousness you would give to a vault full of cash or gold—because in crypto, that is exactly what they represent.

HOW PUBLIC & PRIVATE KEYS WORK

Every cryptocurrency wallet is built on two fundamental cryptographic components: a **public key** and a **private key**. Understanding the difference between them is essential to understanding both how cryptocurrency works—and how it is secured.

PUBLIC KEY (YOUR WALLET ADDRESS) — SAFE TO SHARE

Your public key functions as your wallet address. It is what others use to send cryptocurrency to you.

- It can be safely shared with anyone.
- It works much like an email address—others need it to send you something, but it grants no access or control.
- It is mathematically derived from your private key, but it cannot be reverse-engineered to reveal the private key.

ANALOGY:
Your public key is like your home address. Anyone can mail something to you, but knowing your address does not allow them to enter your house.

PRIVATE KEY (YOUR SECRET KEY) — NEVER SHARE

Your private key is what proves ownership of your cryptocurrency and authorizes transactions.

- It grants full control over the funds associated with your wallet.
- Anyone who possesses your private key can send, move, or drain your cryptocurrency.
- If your private key is stolen or exposed, your funds can be taken instantly—and permanently.

ANALOGY:
Your private key is like the key to your house. Whoever holds it can unlock the door and take whatever is inside.

Note: *None of the products discussed in this book are endorsements or recommendations. They are presented solely for educational purposes to illustrate various tools, concepts, and approaches to network and cybersecurity. Readers are encouraged to conduct their own research and due diligence to determine which products or solutions best meet their individual needs.*

THE CORE PRINCIPLE

Public keys are meant to be shared.
Private keys are meant to be protected.

HOW TO SECURE YOUR SEED PHRASE & PRIVATE KEYS

Your private key and seed phrase are the **single most important security elements** in cryptocurrency ownership. They are not just passwords or recovery tools—they are the ultimate proof of ownership. Lose them, and you lose access to your funds forever. Expose them, and someone else gains full control of your assets.

Because of this, one rule must be treated as absolute:

NEVER STORE YOUR SEED PHRASE OR PRIVATE KEY DIGITALLY.

Any digital copy—whether saved on a computer, phone, cloud storage service, email draft, password manager, or screenshot—introduces unnecessary risk. Malware, phishing attacks, compromised backups, and data breaches all routinely target digital storage. Once a seed phrase is exposed online, the loss of funds is often immediate and irreversible.

The safest approach is physical, offline storage. Writing your seed phrase down by hand ensures it never touches an internet-connected device. For added durability and protection against fire or water damage, many users choose to store their recovery phrase on a metal backup designed specifically for long-term resilience.

Redundancy also matters. A single backup stored in one location creates a single point of failure. Keeping multiple copies in separate, secure locations—such as a home safe and a safe deposit box—helps protect against theft, loss, or environmental damage. These locations should be chosen carefully and disclosed to as few people as possible.

Finally, your seed phrase must never be shared. No legitimate wallet provider, exchange, developer, or customer support agent will ever ask for it. Anyone requesting your private key or recovery phrase—regardless of how convincing they sound—is attempting to steal your cryptocurrency. There are no exceptions to this rule.

In a system where transactions cannot be reversed and ownership is enforced by cryptography alone, protecting your seed phrase is not optional—it is the foundation of everything else. Strong wallets, hardware devices, and network security all exist to defend this one critical secret.

BEWARE OF CRYPTO SCAMS: THE MOST COMMON TRICK

Scammers aggressively target new cryptocurrency users—especially those who are unfamiliar with wallet security or experiencing a problem for the first time. One of the most common and devastating scams follows a predictable pattern.

It typically begins when a user asks for help in a public crypto community such as Discord, Reddit, or Telegram. Within minutes, several "helpful" individuals may reach out via private message, claiming to be moderators, support staff, or experienced users who can resolve the issue quickly. They often sound knowledgeable, reassuring, and urgent.

At some point, they will claim that in order to help, they need your seed phrase or private key—perhaps to "verify" your wallet, "sync" it, or "recover" lost funds. The moment that information is shared, the scam is complete. Within minutes—sometimes seconds—your wallet is recreated on the attacker's device, and all funds are transferred out permanently.

There is no appeal process. There is no recovery. The blockchain simply records the transaction and moves on.

THIS CANNOT BE EMPHASIZED ENOUGH: NEVER SHARE YOUR SEED PHRASE OR PRIVATE KEY WITH ANYONE—UNDER ANY CIRCUMSTANCES.

THE GOLDEN RULE OF CRYPTO SECURITY:

If someone asks for your private key or seed phrase, **they are trying to steal your cryptocurrency.**
No legitimate service, support team, developer, or administrator will ever need this information. There are **no exceptions**.

Understanding and remembering this rule alone can prevent the vast majority of crypto-related losses.

HOW TO AVOID WALLET SCAMS & FAKE WEBSITES

One of the most common ways cryptocurrency is stolen is not through technical exploits, but through deception. Cybercriminals routinely use fake wallet applications, malicious browser extensions, and counterfeit websites to trick users into willingly handing over their private keys or seed phrases. These scams are effective because they imitate legitimate tools so closely that even experienced users can be fooled.

Understanding how these attacks work—and slowing down long enough to verify what you're interacting with—can prevent irreversible losses.

FAKE WALLET PLUG-INS AND APPLICATIONS

A frequent tactic involves malicious browser extensions or mobile apps that closely resemble real cryptocurrency wallets. These fake wallets often copy logos, branding, and interface designs from well-known projects. Once installed, they quietly capture private keys, seed phrases, or transaction approvals and send them to an attacker.

The safest defense is strict source control. Wallet software should only ever be downloaded from **official websites or verified app stores** associated with legitimate projects such as MetaMask, Ledger, or Trezor. Third-party download links, direct messages, and "recommended" extensions should always be treated as suspicious.

FAKE CRYPTO WEBSITES

Another common scam involves fake websites that are nearly indistinguishable from real exchanges, wallets, or decentralized applications. These sites often differ by only a single character in the domain name or use subtle spelling variations that are easy to miss. Victims are usually directed to these sites through phishing emails, social media messages, or search engine ads.

Once on the fake site, users are prompted to log in, connect a wallet, or enter a seed phrase—handing full control of their funds directly to the attacker.

The simplest protection is also one of the most effective: **always verify the website URL**. Bookmark official sites and access them directly rather than clicking links from emails, messages, or ads. If something feels rushed or urgent, that pressure is often part of the scam.

"TECH SUPPORT" AND IMPERSONATION SCAMS

Perhaps the most devastating scams involve impersonation. Attackers pose as wallet support staff, exchange representatives, or community moderators on platforms like Reddit, Discord, or Telegram. They often approach users who have publicly asked for help and offer quick solutions—requesting the seed phrase or private key "to fix the issue."

This is never legitimate.

No real support agent, developer, or administrator will ever need your private key or seed phrase. Anyone who asks for it is attempting to steal your cryptocurrency.

ABSOLUTE RULE:
If someone asks for your seed phrase or private key, the conversation is over. They are a scammer—without exception.

CHAPTER 9 – TYPES OF CRYPTOCURRENCY WALLETS

Now that we understand what a cryptocurrency wallet actually is—a tool for managing cryptographic keys rather than storing coins—it should be mentioned that there are different types of wallets, and not all wallets serve the same purpose. Wallets differ in how they connect to the internet, how keys are protected, and how easily funds can be accessed. These differences are commonly described using terms such as hot and cold wallets, as well as software and hardware wallets. Each category represents a tradeoff between convenience, security, and control.

Rather than relying on a single wallet for every situation, experienced crypto users often utilize a group of wallets, each varying by purpose and used together for different roles. Hot wallets may be used for frequent transactions and everyday activity, while cold wallets are reserved for long-term storage. Software wallets offer flexibility and ease of use, while hardware wallets prioritize isolation and security. Understanding how these wallet types differ—and how they can complement one another—is essential to designing a system that balances usability with protection as your crypto holdings grow.

TYPES OF CRYPTOCURRENCY WALLETS: HOT VS. COLD STORAGE

As previously stated, a cryptocurrency wallet serves as an access point to your digital assets on the blockchain. While it is often compared to a bank account, this comparison only goes so far. A wallet does not hold funds or rely on an institution—instead, it manages the cryptographic keys that grant you control over your cryptocurrency.

The security of a wallet depends almost entirely on **how and where its private keys are stored**. For this reason, wallets are commonly categorized into two broad types:

- Hot Wallets – Wallets that are connected to the internet
- Cold Wallets – Wallets that keep private keys offline

Each type comes with its own balance of convenience, accessibility, and risk. Hot wallets are easier to use for frequent transactions, while cold wallets prioritize security by minimizing exposure to online threats.

Understanding the differences between hot and cold storage—and knowing when to use each—is a critical step in protecting your cryptocurrency. In the sections that follow, we'll explore how these wallets work, their advantages and risks, and best practices for securing your assets effectively.

HOT WALLETS: ALWAYS ONLINE, FAST BUT LESS SECURE

A **hot wallet** is any cryptocurrency wallet that remains connected to the internet. This constant connectivity makes hot wallets fast and convenient, but it also introduces meaningful security risks. Because these wallets interact directly with online services, browsers, and networks, they are naturally more exposed to phishing attempts, malware, and device compromises.

Hot wallets come in several forms. Some are web-based wallets accessed through a browser, often hosted by exchanges or third-party providers. Others are browser extension wallets designed for interacting with decentralized applications. Mobile wallets allow users to manage funds from a phone, while desktop wallets run directly on a personal computer. Although these options vary in usability and features, they all share one critical trait: they are online whenever the device is online.

This is what makes hot wallets attractive—and dangerous.

WHY PEOPLE USE HOT WALLETS

Hot wallets are popular because they remove friction. They allow users to access funds instantly, approve transactions quickly, and interact with exchanges or decentralized platforms without extra hardware or setup. For beginners, they are often the first point of entry into cryptocurrency because they feel familiar and easy to use.

For everyday activity—such as trading, making payments, or interacting with DeFi applications—hot wallets serve a practical role. The problem arises when convenience replaces caution.

THE INHERENT RISK OF HOT WALLETS

Because hot wallets are always online, they exist in the same threat environment as email, browsers, and operating systems. If a device is compromised, the wallet is compromised. If a phishing site tricks a user into signing a malicious transaction, the wallet provides no safety net. Once funds are sent, they are gone.

Unlike traditional financial systems, there is no fraud department to call and no transaction reversal process. A single mistake can result in permanent loss.

For this reason, hot wallets should never be treated as long-term storage solutions. They are best viewed as **spending wallets**, not savings accounts.

USING HOT WALLETS MORE SAFELY

If you choose to use a hot wallet—and most users will at some point—there are ways to reduce risk.

First, disconnect whenever possible. Leaving wallets permanently connected to exchanges or decentralized applications increases exposure unnecessarily. Connecting only when needed takes seconds and significantly limits attack opportunities.

Second, always use strong authentication. App-based two-factor authentication adds an important layer of protection, especially when compared to SMS-based verification, which is vulnerable to SIM swap attacks.

Finally, limit what you store. Only keep small, operational amounts in a hot wallet. Think of it like cash in your pocket—useful, but never where you keep your life savings.

BEST PRACTICE:

Use hot wallets for daily activity and convenience. Use cold wallets for security and long-term storage.

COLD WALLETS: THE ULTIMATE SECURITY SOLUTION

A **cold wallet** is a cryptocurrency wallet that is kept **offline**, meaning it is not continuously connected to the internet. This single characteristic dramatically reduces its exposure to cyber threats and makes cold storage the most secure way to protect cryptocurrency holdings.

Because cold wallets isolate private keys from online environments, they are largely immune to phishing attacks, malware infections, and remote hacking attempts. For this reason, cold wallets are widely regarded as the gold standard for long-term cryptocurrency storage—especially for larger balances.

Cold storage can take several forms. The most common is a **hardware wallet**, a dedicated physical device designed specifically to store private keys securely. Other forms include **paper wallets**, where keys are printed and stored physically, and **air-gapped devices**, such as computers or phones that are never connected to the internet and are used exclusively for crypto storage.

Each of these methods shares the same core principle: **your private keys never touch an online system**.

WHY COLD WALLETS ARE USED

Cold wallets are designed for security first, not convenience. Their primary strength is that accessing funds requires physical possession of the wallet—or access to the recovery seed phrase that created it.

This makes cold wallets ideal for long-term storage and high-value holdings. In fact, most reputable cryptocurrency exchanges store the majority of their customer funds in cold wallets to reduce the risk of large-scale theft.

Key advantages include:

- Private keys remain offline and protected from malware
- Remote attacks are virtually impossible
- Physical access is required to move funds

The Trade-Off: Security vs. Convenience

Cold wallets are not without downsides. Because they are offline by design, they require extra steps to access and use funds. Transactions typically involve connecting a device, verifying details, and manually approving transfers.

There is also the responsibility of proper backup. If a cold wallet device is lost or destroyed and the seed phrase is not securely stored elsewhere, access to funds may be permanently lost. Hardware wallets also involve an upfront cost, typically ranging from $50 to $200.

These trade-offs are intentional. Cold wallets prioritize protection over speed, making them unsuitable for frequent transactions but ideal for safeguarding assets over time.

HOW HARDWARE WALLETS WORK

A **hardware wallet** is a specialized physical device—often resembling a USB drive—that stores private keys in a secure, offline environment. Even when connected to a computer or mobile device, the private keys never leave the hardware wallet itself.

When a transaction is initiated, the unsigned transaction is sent to the hardware wallet. The device then signs the transaction internally using the private key and sends only the signed transaction back to the computer. At no point is the private key exposed to the internet or the operating system.

This design provides a powerful layer of protection. Even if your computer were infected with malware, an attacker would still be unable to move funds without physically interacting with the hardware wallet—usually by pressing a button or confirming the transaction on the device's screen.

For example, if malicious software attempted to initiate a transfer in the background, the transaction would still require manual approval on the hardware wallet itself. This physical verification step stops many attacks before they can succeed.

THE CORE TAKEAWAY

- Hot wallets prioritize convenience.
- Cold wallets prioritize security.

For anyone holding significant cryptocurrency, cold storage is not optional—it is essential. By keeping private keys offline and under your physical control, cold wallets provide the strongest defense available in a decentralized financial system where responsibility rests entirely with the user.

HOT WALLETS VS. COLD WALLETS: WHICH ONE SHOULD YOU USE?

Choosing between a hot wallet and a cold wallet is not an either–or decision—it's a matter of intent and risk management. Each type of wallet serves a different purpose, and understanding when to use each is key to building a secure crypto strategy.

Hot wallets are designed for accessibility. Because they remain connected to the internet, they allow users to quickly send, receive, trade, and interact with decentralized applications. This convenience makes them well suited for everyday use, but it also increases exposure to online threats such as phishing, malware, and compromised devices.

Cold wallets, by contrast, prioritize security over speed. By keeping private keys offline, they dramatically reduce the attack surface available to hackers. Cold storage is best suited for long-term holdings, savings, and assets you do not need to access regularly. The extra steps required to move funds are a feature—not a flaw—because they add meaningful protection.

AT A GLANCE: KEY DIFFERENCES

Feature	Hot Wallets	Cold Wallets
Security Level	⚠️ Lower (Online)	🔒 Higher (Offline)
Ease of Use	✅ Very Easy	❌ Less Convenient
Risk of Hacking	🚨 Higher	🛡️ Very Low
Best Use Case	Daily transactions	Long-term storage
Common Examples	MetaMask, Trust Wallet	Ledger, Trezor

THE SMART APPROACH: USE BOTH

The most effective strategy is not choosing one wallet type—but combining them.

- Use a hot wallet for small, operational amounts you need for trading, payments, or interacting with DeFi.
- Use a cold wallet to store the majority of your holdings safely offline.

This mirrors how people already manage money in the real world: a small amount in a checking account or wallet for daily use, and the bulk of savings kept somewhere far more secure.

BEST PRACTICE:

Only keep what you are willing to risk in a hot wallet. Everything else belongs in cold storage.

By intentionally separating convenience from security, you reduce the chance that a single mistake, phishing attempt, or compromised device can result in catastrophic loss.

WHAT IS A PAPER WALLET?

A paper wallet is an early form of cold storage that involves generating a cryptocurrency private key and public address and recording them on a physical piece of paper—often printed alongside QR codes for convenience. Because the keys are stored offline, paper wallets were once considered a secure way to protect cryptocurrency from online threats.

The basic process works like this: a wallet generator creates a random private key and corresponding public address, the information is printed and stored offline, and funds are sent to that address. As long as the private key remains secret and the paper remains intact, the cryptocurrency associated with that address stays secure on the blockchain.

When the owner later wants to spend the funds, however, the private key must be imported into a software (hot) wallet so that a transaction can be signed and broadcast to the network. At that moment, the key is exposed to an internet-connected environment.

WHY PAPER WALLETS WERE POPULAR:

Before hardware wallets became widely available, paper wallets offered a simple solution to a serious problem: keeping private keys offline. They were inexpensive, did not require specialized devices, and—when generated correctly—were immune to online hacking and malware. For a time, they were widely promoted as the safest way to store cryptocurrency.

THE HIDDEN RISKS OF PAPER WALLETS

While paper wallets technically achieve offline storage, experience has shown that they introduce a different and often more dangerous set of risks. Over time, these drawbacks have made paper wallets far less reliable than they initially appeared.

Paper is fragile. It can fade, tear, burn, or degrade, sometimes without obvious warning. If a paper wallet is damaged or destroyed, access to the funds may be lost permanently. Unlike a hardware wallet, there is no redundancy or recovery mechanism built into the storage method itself.

The process of creating a paper wallet also introduces risk. If the wallet is generated on a computer that is infected with malware—or even one that is connected to the internet at the time—the private key may already be compromised before it is ever printed. Many losses attributed to paper wallets stem from insecure key generation rather than the paper itself.

Perhaps the biggest flaw is usability. To actually spend funds stored in a paper wallet, the private key must be imported into a hot wallet. This immediately exposes the key to the internet and defeats the primary advantage of cold storage. If the key is reused or partially spent incorrectly, it can also lead to accidental loss through address reuse or improper handling.

Finally, paper wallets offer no protection if stolen. Anyone who gains access to the paper automatically gains full control over the funds, with no PIN, passphrase, or secondary verification to stop them.

HARDWARE WALLETS VS. PAPER WALLETS: WHICH IS BETTER?

In theory, both hardware wallets and paper wallets keep private keys offline. In practice, hardware wallets solve many of the real-world problems that paper wallets cannot.

Hardware wallets protect keys inside a secure device, resist physical tampering, require PINs and manual confirmation, and allow transactions to be signed without ever exposing the private key. Paper wallets rely entirely on physical secrecy and perfect handling—something that is difficult to guarantee over time.

Because of these differences, paper wallets are generally considered outdated and risky for most users today.

Feature	Hardware Wallet	Paper Wallet
Security Level	🔒 Very High	⚠ Moderate
Risk of Physical Damage	⚠ Moderate	🚨 High
Ease of Use	✅ Convenient	❌ Difficult
Risk of Theft	⚠ Requires physical access & PIN	🚨 If stolen, no protection
Best For	Long-term storage with strong security	Offline storage (only with extreme precautions)

BEST PRACTICE:
For long-term storage, a reputable hardware wallet provides far stronger security, better usability, and significantly lower risk than a paper wallet.

Paper wallets may still function as cold storage under extreme precautions, but for modern cryptocurrency users—especially those protecting meaningful value—they are no longer the preferred solution.

WHAT IS A SOFTWARE WALLET?

A software wallet is a digital application that allows users to interact with the blockchain in order to send, receive, and manage cryptocurrency. Rather than storing funds themselves, software wallets manage the private keys that grant access to crypto assets recorded on the blockchain.

Software wallets come in several common forms:

✅ **Desktop wallets** – Installed directly on a computer running Windows, macOS, or Linux. These often offer strong control and advanced features, but security depends heavily on the safety of the device itself.
✅ **Mobile wallets** – Applications for iOS and Android devices, designed for convenience and everyday transactions.
✅ **Web-based wallets** – Accessed through a web browser, often provided by exchanges or third-party services. These are convenient but typically rely on external infrastructure.
✅ **Browser extension wallets** – Plug-ins that integrate directly with a web browser, commonly used for interacting with decentralized applications (dApps) and DeFi platforms.

IMPORTANT NOTE:
Many modern software wallets are cross-platform, allowing the same wallet to be accessed from multiple devices using a shared recovery phrase. While this improves convenience, it also increases the importance of securing every device that has access to the wallet.

Software wallets are widely used due to their accessibility and ease of use, but because they remain connected to the internet, they are generally considered **hot wallets** and carry higher exposure to malware, phishing, and device compromise.

WHAT IS A HARDWARE WALLET?

A hardware wallet is a physical device specifically designed to store cryptocurrency private keys in an offline environment. Unlike software wallets, which rely on internet-connected devices, hardware wallets isolate private keys from operating systems, browsers, and networks—dramatically reducing exposure to hacking, malware, and phishing attacks.

At the core of a hardware wallet is a secure microcontroller that generates and stores private keys internally. These keys never leave the device. When a transaction is initiated, the wallet signs it internally and sends only the signed transaction to the connected computer or mobile device. Even if the host system were compromised, an attacker would still be unable to access the private keys or authorize transactions without physical interaction with the hardware wallet itself.

Most hardware wallets also require manual confirmation—such as pressing a button or approving the transaction on a built-in screen—before any funds can be moved. This physical verification step acts as a final safeguard, preventing malicious software from silently draining a wallet in the background.

Many modern hardware wallets use open-source software, allowing independent researchers to audit the code for vulnerabilities. This transparency is an important trust signal in a security-critical device.

SECURITY RECORD

To date, there are essentially no verified cases of cryptocurrency being stolen directly from a properly used, legitimate hardware wallet. When losses do occur, they almost always result from compromised recovery seed phrases, unsafe backups, or physical access combined with poor security practices—not a failure of the hardware wallet itself.

LIMITATIONS & PRECAUTIONS FOR HARDWARE WALLETS

While hardware wallets provide one of the strongest security models available, they are not a magic solution. Their effectiveness depends on how responsibly they are used.

There are three primary risks to be aware of:

- Physical theft – If someone gains access to both the hardware wallet and its recovery seed phrase, they can fully control the funds.
- Supply-chain attacks – Purchasing a wallet from an unofficial or unverified seller introduces the risk of tampering before you ever receive the device.
- Loss or destruction – A lost or damaged hardware wallet is not a problem if the seed phrase is securely backed up. Without that backup, access to funds is permanently lost.

These risks do not negate the value of hardware wallets—but they highlight the importance of proper setup and backup procedures.

BEST PRACTICES FOR USING A HARDWARE WALLET

To get the full security benefit of a hardware wallet, follow these principles:

- Purchase only from the official manufacturer, never from third-party marketplaces
- Set a strong, unique PIN on the device
- Write down the seed phrase by hand and store it securely offline
- Never store the seed phrase digitally or photograph it
- Always verify transaction details on the device itself before approving

SECURITY INSIGHT:

Whenever possible, choose hardware wallets that support open-source firmware. Publicly auditable code reduces the risk of hidden vulnerabilities and increases long-term trust in the device.

THE BOTTOM LINE

Hardware wallets shift crypto security from a software problem to a physical control problem. They dramatically reduce the risk of remote theft by ensuring that private keys remain offline and that transactions require deliberate, physical approval.

Used correctly, a hardware wallet is one of the most powerful tools available for protecting cryptocurrency in a system where ownership is absolute—and mistakes are irreversible.

HARDWARE & PAPER WALLETS: COLD STORAGE SECURITY

Cold storage exists to solve one core problem in cryptocurrency security: how to keep private keys completely isolated from the internet. Hardware wallets and paper wallets both attempt to solve this problem by storing keys offline, but they do so in very different ways—and with very different risk profiles.

A hardware wallet is a purpose-built device designed specifically to generate and store private keys in a secure, offline environment. Unlike software wallets, hardware wallets never expose private keys to the operating system, browser, or internet-connected applications. Even when connected to a computer or mobile device, the signing of transactions happens internally within the hardware wallet itself. This makes hardware wallets one of the most secure and practical cold storage solutions available today.

Paper wallets, by contrast, represent an older approach to cold storage. A paper wallet is simply a physical document containing a wallet's public address and private key—often printed as text or QR codes. While paper wallets technically keep keys offline, they lack many of the safety mechanisms built into modern hardware wallets. Over time, issues such as physical degradation, accidental exposure, insecure key generation, and user error have made paper wallets far less reliable than they initially appeared.

For these reasons, paper wallets have largely fallen out of favor. While they can still function as a form of cold storage, they are generally not recommended for most users today, especially given the availability of secure, user-friendly hardware wallets that dramatically reduce the risk of mistakes.

In the sections that follow, we'll take a closer look at how hardware wallets protect private keys, why they have become the preferred cold storage method, and the specific risks and limitations associated with paper wallets—so you can make informed decisions about how to secure your cryptocurrency for the long term.

HYBRID WALLET STRATEGIES: COMBINING HOT & COLD STORAGE

One of the most effective ways to protect cryptocurrency is to combine different wallet types into a single, intentional workflow. Rather than choosing between security and convenience, hybrid strategies allow you to benefit from both—limiting risk while maintaining usability.

Two hybrid approaches stand out as particularly effective: air-gapped wallets for maximum isolation, and hardware wallets paired with hot wallets for secure everyday use.

AIR-GAPPED WALLETS: MAXIMUM ISOLATION, MINIMAL EXPOSURE

An air-gapped wallet is a form of cold storage designed to remain completely isolated from the internet. Unlike traditional hardware wallets that connect via USB or Bluetooth, air-gapped wallets communicate exclusively through QR codes. This design ensures that private keys never come into contact with an internet-connected device—eliminating entire classes of remote attacks.

In practice, an air-gapped transaction works as a coordinated exchange between two devices. A transaction is prepared on a software wallet, but not signed. That unsigned transaction is displayed as a QR code, which the air-gapped wallet scans using its built-in camera. The air-gapped device then signs the transaction internally and displays a new QR code containing the signed transaction. That code is scanned back into the software wallet and broadcast to the blockchain.

At no point does the private key leave the air-gapped device, nor does the device ever need to connect to the internet.

WHY THIS MATTERS:

Because the signing process occurs entirely offline, malware, browser exploits, and network-based attacks have no path to access the private keys. This makes air-gapped wallets one of the most secure options available for high-value storage.

COMBINING A HOT WALLET WITH A HARDWARE WALLET

A more common—and very practical—hybrid strategy is to pair a hot software wallet with a hardware wallet. In this setup, the software wallet provides the interface for interacting with decentralized applications and exchanges, while the hardware wallet retains full control over the private keys.

For example, a browser-based wallet such as MetaMask can be connected to a hardware wallet like Ledger. The software wallet handles connectivity and usability, but every transaction must be physically approved on the hardware device.

In a decentralized exchange swap, this means the software wallet prepares the transaction and requests approval, but the hardware wallet displays the transaction details—amounts, destination, and fees—on its own screen. The user must then manually confirm the transaction on the device before it can be signed and submitted to the blockchain.

WHY THIS WORKS:
Even if the software wallet, browser, or computer is compromised, an attacker cannot move funds without physical access to the hardware wallet. The requirement for manual approval acts as a final, powerful safeguard against unauthorized transfers.

THE CORE ADVANTAGE OF HYBRID STRATEGIES

Hybrid wallet strategies are effective because they separate responsibilities:

- Software wallets handle interaction and convenience
- Hardware or air-gapped wallets handle security and key ownership

This separation ensures that no single failure—whether malware, phishing, or device compromise—can result in total loss. Instead of relying on one tool, you create a system where multiple layers must fail before funds are at risk.

In a decentralized financial system where mistakes are irreversible, this kind of intentional design is not overkill—it's responsible ownership.

HYBRID WALLET STRATEGIES & SECURE STORAGE SOLUTIONS

Securing cryptocurrency is not about relying on a single wallet or tool—it's about **designing a system**. Just as strong cybersecurity relies on layered defenses, effective crypto security comes from combining multiple wallet types and storage practices in a way that limits exposure and contains risk.

A hybrid wallet strategy deliberately separates convenience from security. Hot wallets, software wallets, and browser-based tools provide speed and accessibility, while hardware wallets and other forms of cold storage protect long-term holdings by keeping private keys offline. When used together, these tools create a defense-in-depth approach that reduces the chance that a single mistake, phishing attempt, or compromised device can result in total loss.

Rather than placing all assets in one location, a hybrid approach allows you to segment your holdings. Smaller balances can be kept accessible for everyday use, while the majority of funds remain isolated and protected. This mirrors how people manage physical money—carrying what they need while keeping savings secured elsewhere.

In the sections that follow, we'll explore practical hybrid wallet strategies, examine best practices for storing private keys and seed phrases safely, and break down the most common threats facing crypto users—along with clear, actionable steps to defend against them.

This is where individual tools come together into a cohesive security strategy—one designed to protect your assets in a system where responsibility rests entirely with you.

HOW TO AVOID WALLET SCAMS & FAKE WEBSITES

One of the most common ways cryptocurrency is stolen is not through technical exploits, but through deception. Cybercriminals routinely use fake wallet applications, malicious browser extensions, and counterfeit websites to trick users into willingly handing over their private keys or seed phrases. These scams are effective because they imitate legitimate tools so closely that even experienced users can be fooled.

Understanding how these attacks work—and slowing down long enough to verify what you're interacting with—can prevent irreversible losses.

FAKE WALLET PLUG-INS AND APPLICATIONS

A frequent tactic involves malicious browser extensions or mobile apps that closely resemble real cryptocurrency wallets. These fake wallets often copy logos, branding, and interface designs from well-known projects. Once installed, they quietly capture private keys, seed phrases, or transaction approvals and send them to an attacker.

The safest defense is strict source control. Wallet software should only ever be downloaded from **official websites or verified app stores** associated with legitimate projects such as MetaMask, Ledger, or Trezor. Third-party download links, direct messages, and "recommended" extensions should always be treated as suspicious.

FAKE CRYPTO WEBSITES

Another common scam involves fake websites that are nearly indistinguishable from real exchanges, wallets, or decentralized applications. These sites often differ by only a single character in the domain name or use subtle spelling variations that are easy to miss. Victims are usually directed to these sites through phishing emails, social media messages, or search engine ads.

Once on the fake site, users are prompted to log in, connect a wallet, or enter a seed phrase—handing full control of their funds directly to the attacker.

The simplest protection is also one of the most effective: **always verify the website URL**. Bookmark official sites and access them directly rather than clicking links from emails, messages, or ads. If something feels rushed or urgent, that pressure is often part of the scam.

"TECH SUPPORT" AND IMPERSONATION SCAMS

Perhaps the most devastating scams involve impersonation. Attackers pose as wallet support staff, exchange representatives, or community moderators on platforms like Reddit, Discord, or Telegram. They often approach users who have publicly asked for help and offer quick solutions—requesting the seed phrase or private key "to fix the issue."

This is never legitimate.

No real support agent, developer, or administrator will ever need your private key or seed phrase. Anyone who asks for it is attempting to steal your cryptocurrency.

ABSOLUTE RULE:

If someone asks for your seed phrase or private key, the conversation is over. They are a scammer—without exception.

THE CORE DEFENSE: SLOW DOWN AND VERIFY

Wallet scams succeed because they exploit urgency, familiarity, and trust. The most powerful defense is not advanced technology—it's patience. Take the extra moment to verify sources, double-check URLs, and question unsolicited help.

In a system where transactions are irreversible and ownership is absolute, caution is not paranoia—it's responsibility.

BEST PRACTICES FOR SECURING DIFFERENT WALLET TYPES

No single wallet type is inherently "safe" or "unsafe." Security comes from how a wallet is used, how much value it holds, and how well it is integrated into a broader strategy. Each wallet type introduces different risks, and each requires a slightly different approach to protection.

Securing Software Wallets (Hot Wallets)

Software wallets are designed for accessibility, which makes them ideal for frequent transactions—but also more exposed to attack. The most important principle when using a hot wallet is limitation: limit both access and stored value.

Choose wallets with a strong security history and active development. Protect access with a complex, unique password stored in an encrypted password vault rather than memorized or reused. Where available, enable two-factor authentication to reduce the risk of unauthorized access if credentials are compromised.

Equally important is recovery planning. Your seed phrase should be written down and stored offline in a secure location. For long-term durability, many users opt for metal seed storage rather than paper, which can degrade over time.

Hot wallets should be treated as operational tools—not vaults.

SECURING BROWSER PLUG-IN WALLETS

Browser extension wallets introduce an additional risk surface because they operate inside the browser environment, which is a frequent target for phishing, malicious scripts, and fake extensions.

Always disconnect wallet sessions after completing transactions, and enable automatic locking after short periods of inactivity. Keeping the wallet software updated is critical, as security vulnerabilities are regularly discovered and patched.

Perhaps most importantly, only install browser wallets from verified, official sources. Fake extensions that mimic legitimate wallets—such as MetaMask—are one of the most common methods attackers use to steal funds.

SECURING HARDWARE WALLETS (COLD STORAGE)

Hardware wallets offer strong protection by keeping private keys offline, but they still require responsible handling. A strong PIN should be set on the device, avoiding anything easily guessed or personally identifiable.

Your recovery seed phrase should never be stored with the hardware wallet itself. The device and the seed should be physically separated so that losing one does not compromise the other. Treat the hardware wallet as you would physical valuables—store it in a secure location and disconnect it when not in use.

A hardware wallet's security advantage comes from isolation. Preserving that isolation is key.

SECURING PAPER WALLETS

Paper wallets require extreme care and discipline. If used at all, they should be generated on a clean, offline system and printed in a way that minimizes degradation. Laser printing is preferable, as inkjet prints can smudge or fade over time.

Paper wallets should be protected against physical damage through lamination or sealed storage, and multiple copies should be kept in separate secure locations. Under no circumstances should digital copies be created—no photos, screenshots, scans, or cloud backups.

Because paper wallets lack safeguards like PINs, confirmations, or recovery mechanisms, they should be viewed as fragile and high-risk storage methods.

FINAL THOUGHTS: BUILDING A SECURE WALLET STRATEGY

There is no single "best" wallet—only **appropriate use cases**. A strong crypto security posture comes from combining tools intentionally rather than relying on one solution for everything.

Hot wallets are best reserved for small amounts needed for convenience and daily use. Long-term holdings belong in cold storage, preferably on a hardware wallet or air-gapped setup. Seed phrases should be backed up offline in multiple secure locations, and every website or application interacting with your crypto should be verified before use.

By applying a layered, defense-in-depth approach—separating convenience from security and reducing single points of failure—you dramatically reduce the risk of theft, compromise, or accidental loss. In a system where ownership is absolute and mistakes are permanent, this kind of planning isn't excessive—it's essential.

THE BEST WAY TO STORE CRYPTO SAFELY AND CHOOSING THE RIGHT WALLET FOR YOUR NEEDS

Storing cryptocurrency securely is not about finding a single perfect tool—it's about matching the right storage method to your goals, risk tolerance, and usage habits. That said, some options clearly provide stronger protection than others.

For long-term security, a reputable hardware wallet remains the best solution available. By keeping private keys offline and requiring physical confirmation for transactions, hardware wallets dramatically reduce the risk of remote theft and accidental loss. For anyone holding meaningful value in cryptocurrency, this is the foundation of a responsible security strategy.

Paper wallets, while technically capable of offline storage, demand extreme caution. They lack built-in safeguards, are vulnerable to physical damage or theft, and introduce risk when funds are eventually spent. While they can work under carefully controlled conditions, they are no longer the preferred choice for most users.

Hot wallets still serve an important role, but only when used intentionally. They are best reserved for small amounts needed for daily transactions, trading, or interaction with decentralized applications. Keeping large balances in an always-online wallet unnecessarily exposes funds to avoidable risk.

By understanding the strengths and weaknesses of each wallet type—and by separating convenience from security—you place yourself in control rather than at the mercy of chance. In a decentralized financial system, thoughtful wallet selection is one of the most important decisions you can make.

Choosing the right wallet is ultimately about understanding how you use cryptocurrency and aligning your security practices with that reality. There is no single "perfect" wallet—only tools that are appropriate for different purposes.

If you trade frequently or interact with decentralized applications, a hot wallet may be necessary for convenience. In that case, security becomes non-negotiable. Strong passwords, two-factor authentication, and careful browsing habits are essential to reduce risk. Hot wallets should be treated like spending accounts, not long-term vaults.

If you hold significant amounts of cryptocurrency, long-term security should take priority over ease of access. A hardware wallet provides the strongest protection by keeping private keys offline and under your direct physical control. For serious holders, this is not an optional upgrade—it is a foundational safeguard.

One mistake many users make is leaving assets on centralized exchanges. While exchanges are convenient, they are custodial by nature, meaning you do not control the private keys. History has repeatedly shown that exchanges can be hacked, frozen, or shut down without warning. If you do not control the keys, you do not truly own the crypto.

Regardless of which wallet you choose, backing up your seed phrase securely is critical. A wallet can always be replaced; access to your seed phrase cannot. Stored properly, it is your ultimate recovery tool. Lost or exposed, it becomes the fastest path to permanent loss.

By understanding the trade-offs between convenience and security—and by intentionally choosing where each type of wallet fits into your strategy—you dramatically reduce your exposure to risk. In a system built on personal responsibility, informed wallet choices are one of the most important steps you can take to protect your cryptocurrency.

NEXT: HOW TO SECURE YOUR WALLET FROM THREATS

Now that you understand the different types of wallets and how they store your keys, the next section will focus on protecting those wallets from real-world threats, including:

- How hackers and phishing scams target wallet users
- Best practices for safeguarding private keys and recovery phrases
- What to do if a wallet is lost, stolen, or compromised

This is where theory meets practice—and where strong habits make the difference between security and loss.

CHAPTER 10 - WHERE DO YOU GET CRYPTOCURRENCY?

CRYPTOCURRENCY EXCHANGES: CENTRALIZED VS. DECENTRALIZED

If you want to buy, sell, or trade cryptocurrency, you will typically do so through a cryptocurrency exchange. Much like a stock exchange, these platforms provide the infrastructure to convert one asset into another and to execute trades efficiently. For most people, exchanges are the primary entry point into crypto, especially because you generally cannot walk into a local bank and purchase digital assets directly.

Not all exchanges operate the same way, and the differences matter—especially when security and custody are involved. When evaluating an exchange, it helps to consider a few practical factors: which cryptocurrencies and networks are supported, how deposits and withdrawals work (fiat vs. crypto), and what security controls are available such as two-factor authentication and withdrawal protections. Some platforms also claim to store portions of customer funds in cold storage or offer limited forms of insurance, though the details and limitations vary widely.

At a high level, exchanges fall into two main categories:

- **Centralized exchanges (CEX)** are operated by a company that acts as an intermediary. These platforms typically offer user-friendly interfaces, customer support, and direct fiat on-ramps, but they also introduce custodial risk because the platform may hold assets on the user's behalf.
- **Decentralized exchanges (DEX)** operate through smart contracts and allow users to trade directly from their own wallets. This model reduces reliance on intermediaries and increases self-custody, but it also shifts responsibility to the user and often provides little or no customer support if something goes wrong.

Understanding the difference between centralized and decentralized exchanges is critical, because it determines who controls the keys, how transactions are executed, and what recourse—if any—exists when problems occur.

CENTRALIZED EXCHANGES (CEFI): CONVENIENCE & SECURITY TRADE-OFFS

Before decentralized finance (DeFi) became widely accessible, centralized finance—commonly referred to as CeFi—was the primary way most people bought, sold, and traded cryptocurrency. In this model, all transactions flow through a centralized exchange operated by a company that acts as an intermediary between buyers and sellers.

CeFi exchanges function in a way that will feel familiar to anyone who has used traditional financial platforms like E*TRADE or Fidelity. The exchange provides the interface, executes trades, and—most importantly—holds custody of user funds. This means the platform controls the private keys associated with customer wallets, not the individual user.

For many newcomers, this arrangement is appealing. Centralized exchanges abstract away much of the technical complexity involved in managing wallets, private keys, and blockchain interactions. Users can log in with a username and password, reset access if credentials are lost, and rely on customer support when problems arise. Some platforms also advertise insurance coverage for certain custodial holdings, further reinforcing a sense of security.

However, this convenience comes with a significant trade-off.

THE BIG CEFI TRADE-OFF: CONVENIENCE VS. CONTROL

A common saying in the crypto world captures this reality succinctly:

"NOT YOUR KEYS, NOT YOUR CRYPTO."

When you use a centralized exchange, you do not actually control your cryptocurrency at the protocol level. The exchange does. Your account balance represents a claim on assets held by the platform, not direct ownership enforced by cryptographic keys under your control.

This model offers real advantages:

- Centralized exchanges are generally easy to use and well suited for beginners.
- They provide direct fiat on-ramps, allowing users to purchase crypto with traditional currencies like USD or EUR.
- Transactions are typically fast and inexpensive compared to on-chain decentralized alternatives.

- Customer support exists to assist with account access, errors, or disputes.

At the same time, the risks are equally real:

- You do not control your private keys, which means ultimate ownership rests with the exchange.
- Exchanges are attractive targets for hackers, and history has shown that even large platforms can be compromised.
- Accounts can be frozen, restricted, or locked during periods of high volatility, regulatory pressure, or internal policy changes—sometimes without clear timelines or recourse.

In CeFi, you are trusting a third party not only to safeguard your assets, but also to allow you access to them when you want or need it.

KYC & REGULATION: THE COST OF CENTRALIZATION

Most centralized exchanges are subject to regulatory requirements, which typically include Know Your Customer (KYC) procedures. KYC requires users to submit personal information such as government-issued identification, proof of address, and sometimes biometric verification. These measures are intended to comply with anti-money-laundering laws and tax reporting obligations.

While regulation can add legitimacy and consumer protections in some contexts, it also introduces new risks that do not exist in fully decentralized systems. Accounts can be flagged, frozen, or restricted based on regulatory changes, jurisdictional issues, or internal compliance decisions—often without immediate resolution.

A well-known example involves Binance. Although Binance operates globally, regulatory pressure in the United States led to the creation of a separate platform, Binance.US, with fewer supported assets and features. Users' access depends not only on the platform itself, but also on their country of residence and evolving regulatory frameworks.

WHY THIS MATTERS

When you use a centralized exchange, you are opting into a system where access to your funds can be influenced by laws, politics, corporate policies, or technical failures—factors entirely outside your control. In extreme cases, funds can remain inaccessible indefinitely.

CeFi exchanges are not inherently bad. They serve an important role in onboarding users and providing liquidity to the market. But they should be understood for what they are: **custodial services**, not true self-custody solutions.

Understanding this distinction is essential before deciding how—and where—you acquire and store your cryptocurrency. In the next section, we'll contrast this model with decentralized exchanges, where users trade directly from their own wallets and control remains firmly in their hands.

POPULAR CENTRALIZED EXCHANGES (CEFI)

Note: *None of the products discussed in this book are endorsements or recommendations. They are presented solely for educational purposes to illustrate various tools, concepts, and approaches to network and cybersecurity. Readers are encouraged to conduct their own research and due diligence to determine which products or solutions best meet their individual needs.*

Some well-known centralized exchanges include:

- Binance – Still the largest global exchange by trading volume and user count, offering hundreds of coins and a wide range of trading products.
- Coinbase – A major U.S.-based exchange known for its beginner-friendly interface, regulatory compliance, and public listing.
- Kraken – One of the longest-running exchanges, known for strong security and a large selection of crypto assets.
- Gemini – A U.S.-based exchange with a focus on compliance and security, also expanding its regulated footprint globally.

KEY TAKEAWAY:

Centralized exchanges prioritize convenience and accessibility, especially for new users. However, they require you to trust a third party with custody of your assets. If security and self-sovereignty are your highest priorities, centralized platforms should be viewed as on-ramps—not long-term storage solutions.

DECENTRALIZED EXCHANGES (DEFI): FULL CONTROL, FULL RESPONSIBILITY

Decentralized exchanges operate on an entirely different model than CeFi platforms. Instead of relying on a company or intermediary, DeFi exchanges use smart contracts on a blockchain to facilitate peer-to-peer trading.

Key characteristics of decentralized exchanges include:

- No central authority – Trades are executed directly between users via smart contracts.
- No KYC requirements – Users are not required to submit personal information or identification.
- User-controlled custody – You retain full control of your private keys; the exchange never holds your funds.

CORE DIFFERENCE:
In DeFi, there is no safety net. You are not trusting an institution—you are trusting code, cryptography, and your own security practices.

This shift represents the true ethos of cryptocurrency: self-custody, personal responsibility, and financial sovereignty. With that freedom comes risk, and understanding both sides of the trade-off is essential before deciding where and how to acquire crypto.

HOW DEFI WORKS

Decentralized Finance (DeFi) exchanges operate directly on blockchain networks using decentralized applications (dApps) and smart contracts rather than centralized intermediaries. Instead of depositing funds into an exchange account, users connect their own wallets and trade directly from them. At no point does the exchange take custody of assets or control private keys.

When a trade is made on a DeFi exchange, a smart contract executes the transaction according to predefined rules. Liquidity is often provided by other users rather than by the exchange itself, and prices are determined algorithmically rather than through traditional order books. This architecture removes intermediaries—but it also removes safety nets.

ADVANTAGES OF DEFI EXCHANGES

DeFi's appeal lies in autonomy and permissionless access:

- Full control over your funds – Assets remain in your wallet at all times; there is no third-party custodian.
- No account restrictions – Anyone with a compatible wallet can trade, regardless of location or identity.
- Reduced reliance on centralized fees – There are no exchange custody or withdrawal fees, only blockchain transaction (gas) fees.

- Broader asset access – DeFi often supports new or experimental tokens long before they appear on centralized exchanges.

This model aligns closely with the original philosophy of cryptocurrency: self-custody, transparency, and resistance to censorship.

RISKS AND TRADE-OFFS OF DEFI

With greater control comes greater responsibility. DeFi removes intermediaries—but also removes recourse.

- No customer support – Mistakes are final. Sending funds to the wrong address or approving a malicious contract cannot be undone.
- Smart contract risk – DeFi platforms rely on code. If that code contains vulnerabilities, funds can be lost even if users act correctly.
- Higher learning curve – Users must understand wallets, gas fees, approvals, and contract interactions to operate safely.

Unlike CeFi, where an exchange may freeze withdrawals or reverse errors (for better or worse), DeFi treats every transaction as final.

KEY TAKEAWAY

DeFi offers maximum autonomy—but zero forgiveness.

You gain full control over your funds and freedom from centralized oversight, but you also assume complete responsibility for security, accuracy, and decision-making. DeFi is not inherently more dangerous than CeFi—but it demands discipline, education, and caution.

In short, DeFi doesn't protect you from mistakes. It gives you the freedom to make them.

COMPARING CEFI VS. DEFI EXCHANGES

At a high level, the difference between centralized and decentralized exchanges comes down to custody, control, and responsibility. Both models allow users to trade cryptocurrency—but they do so in fundamentally different ways, with different trade-offs.

The table below highlights the most important distinctions:

Feature	**CeFi (Centralized Exchange)**	**DeFi (Decentralized Exchange)**
Custody of Funds	Exchange controls funds	You control funds
Ease of Use	✅ Beginner-friendly	⚠️ More complex
KYC / ID Verification	✅ Required	❌ Not required
Customer Support	✅ Available	❌ No support
Primary Security Risk	🚨 Exchange hacks	🚨 Smart contract vulnerabilities
Trading Fees	Generally higher	Typically lower
Examples	Coinbase, Binance, Kraken	Uniswap, PancakeSwap, SushiSwap

CeFi exchanges prioritize accessibility and convenience. They abstract away much of the complexity of wallets, private keys, and blockchain interactions, making them ideal for onboarding new users. However, this convenience comes at the cost of control—you are trusting a third party to safeguard your assets.

DeFi exchanges remove that intermediary entirely. Trades occur directly from your wallet through smart contracts, and you never relinquish custody of your funds. This aligns more closely with the original philosophy of cryptocurrency—but it also means there is no safety net if something goes wrong.

GAS FEES & TRANSACTION COSTS IN DEFI

While DeFi often eliminates exchange-related fees, it introduces another cost: blockchain transaction fees, commonly referred to as gas fees. These fees are paid to miners or validators for processing transactions and can vary dramatically depending on the network.

On Ethereum, gas fees fluctuate based on network congestion and demand. During peak periods, a single transaction can cost anywhere from $100 to $400—or more—making small trades impractical. Other networks were designed specifically to address this issue.

For example:

- Binance Smart Chain and Avalanche typically offer transaction fees around .1% of the transaction
- These lower fees have made them popular alternatives for everyday DeFi activity.
- Ethereum continues to undergo upgrades intended to improve scalability and reduce fees, though these improvements are incremental rather than immediate.

KEY TAKEAWAY:

DeFi is often cheaper on newer or less congested blockchains. Ethereum remains powerful and dominant—but costly during high demand.

FINAL THOUGHTS: CHOOSING THE RIGHT EXCHANGE FOR YOU

There is no universally "correct" choice between CeFi and DeFi—only choices that make sense for your goals, experience level, and risk tolerance.

- If you value full control and censorship resistance, DeFi exchanges paired with a hardware wallet offer the strongest self-custody model.
- If you prioritize ease of use and fiat on-ramps, CeFi exchanges provide a smoother entry point—but large balances should be withdrawn to a wallet you control.
- If you trade frequently, many users adopt a hybrid approach, using CeFi for fiat access and liquidity while relying on DeFi for long-term control and flexibility.

No matter which model you choose, one rule remains constant: Strong security habits matter more than the platform itself. Exchanges—centralized or decentralized—are tools. How safely you use them determines whether they empower you or expose you.

CHAPTER 11 - CRYPTO SECURITY – SCAMS, HACKS, AND RISK MITIGATION

CRYPTO SECURITY IS NOT OPTIONAL

Cryptocurrency gives you something traditional finance never did: **direct ownership**.
There is no bank manager, no fraud department, no chargeback system standing between you and your assets. That freedom is powerful—but it also means responsibility cannot be outsourced.

If you are new to crypto, centralized exchanges (CeFi) are usually the safest starting point. They offer familiar interfaces, customer support, and basic protections. But as soon as you step beyond major assets like Bitcoin and Ethereum—or move into decentralized finance (DeFi)—you enter a very different environment.

In DeFi, **you are your own bank**. There is no safety net.

This chapter is about understanding how crypto theft actually happens, why scams are so effective, and how to build a security posture that makes you a difficult target. Not invulnerable—just expensive and time-consuming to attack.

Most losses in crypto do not come from brilliant hackers.
They come from **avoidable mistakes**.

THE DEFENSE-IN-DEPTH MINDSET

Before examining specific scams, it's important to understand the core security philosophy that protects experienced users.

Defense-in-depth means **layered protection**. No single tool or habit keeps you safe. Instead, security works by forcing an attacker to defeat multiple barriers—each one increasing the effort, risk, and time required.

Hackers and scammers almost never pursue the hardest target. They go after **low-hanging fruit**:

- Users who store seed phrases digitally
- Wallets left permanently connected to sites
- Weak passwords or no two-factor authentication
- People who rush or panic

Your goal is simple: **stop being easy**.

With that framework in mind, let's examine the most common threat categories.

CENTRALIZED EXCHANGE (CEFI) THREATS

Centralized exchanges act as custodians. They hold your funds and manage your private keys. This creates convenience—but also concentrates risk.

FAKE OR UNTRUSTWORTHY EXCHANGES

Some platforms are designed from the start to steal deposits. Others operate legitimately at first, then restrict withdrawals or disappear.

These exchanges often display:

- Poor or nonexistent third-party reviews
- Excessive withdrawal fees or delays
- Promises of guaranteed returns

The safest approach is boring but effective: use established platforms, verify URLs manually, and research before depositing funds.

SOCIAL ENGINEERING AND FAKE SUPPORT

One of the most common CeFi scams is impersonation.

It usually begins when a user asks for help publicly—on Discord, Reddit, or Telegram. Shortly after, a "support agent" sends a private message offering fast assistance. The attacker sounds knowledgeable, reassuring, and urgent.

At some point, they ask for login credentials, verification codes, or recovery information.

The moment anything sensitive is shared, the account is drained.

Legitimate exchanges **do not initiate private messages**. Support is always handled through official channels.

PHISHING EMAILS AND WEBSITES

Phishing remains one of the most effective attack methods because it exploits panic.

An email claims your account has been compromised. A link leads to a site that looks identical to the real exchange. You log in—and unknowingly hand your credentials to an attacker.

The defense is simple but non-negotiable:

- Never click links in unsolicited messages
- Access exchanges through bookmarks or manually typed URLs
- Let password managers warn you when a site is suspicious

FAKE GIVEAWAYS

Any offer that requires sending crypto to receive more crypto is a scam. There are no exceptions. These are extremely common, and if you're in the Defi space for long, you will certainly accumulate your share of fake tokens.

Fake giveaway, or airdrops, exploit the idea of "free crypto" to trick users into **authorizing theft**, not sending funds directly.

A fake airdrop typically appears as:

- An unexpected token appearing in your wallet
- A social media post or email promising "free tokens"
- A website claiming you must "claim," "verify," or "activate" the airdrop

The scam works when the user is directed to a website that asks them to:

- Connect their wallet
- Approve a transaction
- Sign a message or contract

Once approved, the malicious contract is granted permission to **drain tokens from the wallet**—sometimes immediately, sometimes later.

Important:
In most fake airdrop scams, the victim never sends crypto directly. They lose funds because they **authorized access**.

DECENTRALIZED EXCHANGE (DEFI) THREATS

DeFi removes intermediaries—but it also removes safety rails. You control your funds directly, which means mistakes are final.

RUG PULLS

A rug pull occurs when developers drain liquidity from a project, leaving investors with worthless tokens.

These scams rely on:

- Hype and urgency
- Anonymous or unverifiable teams
- Unlocked liquidity pools
- Complex or misleading token mechanics

Once liquidity is gone, selling becomes impossible.

PUMP-AND-DUMP SCHEMES

Pump-and-dump schemes artificially inflate a token's price through coordinated hype. Early participants sell into the surge, while late buyers absorb the losses.

High concentration of tokens among a few wallets is often a warning sign.

TROJAN TOKENS AND AIRDROP TRAPS

Sometimes a token appears in your wallet without explanation. Interacting with it—attempting to trade or transfer—can grant malicious permissions that drain your wallet.

Unknown tokens should be ignored entirely.
They cost nothing to leave untouched.

FAKE DEFI WEBSITES

Attackers clone legitimate DeFi platforms and trick users into approving malicious transactions.

Search ads and typo-squatting domains are common entry points.

Manual URL entry, bookmarks, and hardware wallets dramatically reduce risk.

CASE STUDY: THE SQUID TOKEN RUG PULL

The SQUID token scam demonstrates how sophisticated modern rug pulls can be.

The project appeared legitimate:

- A polished website
- A detailed whitepaper
- Social media presence
- A supposed play-to-earn game

The token price surged from pennies to over $2,800.

What investors missed was a hidden restriction: selling SQUID required purchasing another token ("Marble") in escalating ratios. Most holders could never exit.

When liquidity peaked, the developers drained the pool.
The price collapsed within minutes.

Losses totaled approximately $10–11 million.

The lesson is not "don't trust anything."
It's **read token mechanics carefully** and ensure exits are actually possible.

SECURITY IS ABOUT FRICTION

Every successful crypto theft follows the same pattern:

- Reduce friction
- Create urgency
- Exploit trust

Defense-in-depth does the opposite. It adds friction—just enough that attackers move on.

This is why:

- Hardware wallets matter
- Manual transaction approval matters
- Disconnecting wallets matters
- Slowing down matters

Security is not paranoia.
It is discipline.

FINAL PRINCIPLES OF CRYPTO SECURITY

- Never share private keys or seed phrases—ever
- Assume transactions cannot be reversed
- Treat convenience as risk
- Verify everything before interacting
- Store long-term assets offline

Crypto rewards responsibility and punishes complacency.

Crypto does not forgive mistakes—but it **does reward preparation**. You do not need to be perfect. You only need to be harder to exploit than the next target.

CHAPTER 12 - EXAMPLES OF CRYPTO STRATEGIES

SECURITY IS A SPECTRUM, NOT A SWITCH

There is no single "correct" way to secure cryptocurrency. The right strategy depends on one simple factor: how much value you are protecting.

Someone holding a few hundred dollars in crypto does not need the same setup as someone managing six figures across multiple wallets and platforms. What matters is alignment—your security posture should scale with the size and importance of your holdings.

As a general rule, once your crypto holdings exceed $10,000, investing in security tools such as a hardware wallet, VPN, and secure communications becomes less of an expense and more of an insurance policy. The cost of prevention is small compared to the cost of a single mistake.

The following examples illustrate three common security tiers: Good, Better, and Best. Each level builds on the previous one, adding layers of protection as value and exposure increase.

GOOD: BASIC SECURITY FOR CASUAL INVESTORS

This level is appropriate for beginners and casual investors who primarily use centralized exchanges and are holding relatively modest amounts of cryptocurrency.

At this stage, the goal is to eliminate the most common and easily exploited weaknesses.

A solid "Good" setup includes:

- Strong, unique passwords for exchange and wallet accounts
- Two-factor authentication (2FA) enabled for logins and withdrawals
- Use of a reputable centralized exchange
- An offline backup of your wallet seed phrase (never stored in the cloud)
- Keeping wallet and device software up to date

This level of security protects against many everyday threats such as account takeovers, password reuse attacks, and basic phishing attempts.

The limitation:
Your assets are still custodial. If the exchange is hacked, frozen, or restricts withdrawals, your access to funds depends entirely on a third party.

This is acceptable for learning and light exposure—but it is not where long-term value should live.

BETTER: INTERMEDIATE SECURITY FOR ACTIVE TRADERS AND DEFI USERS

As soon as you begin interacting with DeFi, holding meaningful balances, or moving funds frequently, the security model must change. Control shifts from the exchange to you—and your defenses must follow.

At the "Better" level, users begin separating access from storage.

A stronger setup includes:

- Withdrawing assets from exchanges into a hardware wallet
- Using a password manager to generate and store unique credentials
- Connecting hot wallets only when needed and disconnecting after use
- Using a VPN when accessing DeFi platforms or managing wallets
- Enabling browser protections and blocking malicious scripts
- Manually verifying URLs instead of clicking ads or links

This tier dramatically reduces the risk of catastrophic loss from exchange failures, phishing, or compromised credentials.

THE REMAINING RISK:
If your seed phrase or private keys are exposed—through poor backups, social engineering, or malware—funds can still be stolen. Responsibility has increased, and so has the need for discipline.

BEST: ADVANCED SECURITY FOR HIGH-VALUE HOLDINGS

This level is designed for users managing large balances, long-term holdings, or assets they cannot afford to lose. Here, security is treated as a system—not a tool.

At this stage, the priority is eliminating single points of failure.

A "Best" setup includes everything from the previous tiers, plus:

- Using a hardware wallet for *all* meaningful transactions
- Never storing significant balances in hot wallets

- Layered two-factor authentication for both access and withdrawals
- Multiple, geographically separated seed phrase backups
- Fireproof and durable seed storage (such as metal backups)
- Air-gapped wallets for maximum isolation
- A dedicated device for managing high-value wallets
- Strict transaction verification on the hardware wallet screen

This approach introduces friction—and that is intentional. Every additional step creates another barrier an attacker must overcome.

THE TRADE-OFF:
This setup requires more time, care, and organization. But it reduces risk to the lowest practical level available in a self-custody system.

CHOOSING THE RIGHT LEVEL FOR YOU

Security is not about paranoia—it is about proportional response.

- Small holdings justify simple, clean setups
- Active trading demands stronger controls
- Significant wealth requires layered defenses

What matters most is consistency. A modest setup followed carefully is safer than an advanced setup used carelessly.

Crypto does not punish ignorance—it punishes assumptions.

By matching your security strategy to the value you are protecting, you move from reactive defense to intentional ownership. And in a system where transactions are irreversible and responsibility cannot be outsourced, that mindset is the most powerful security tool you have.

FINAL THOUGHTS: THE IMPORTANCE OF SECURITY MINDSET

The most important shift a crypto holder can make is not technological—it is psychological.

You must begin thinking like a bank.

A bank does not rely on a single lock or a single guard. It uses layered defenses designed to slow attackers down, expose mistakes, and make theft more effort than it is worth. Vaults protect assets. Guards verify access. Surveillance detects abnormal behavior. Each layer exists because no single safeguard is perfect.

Crypto security works the same way.

Cold wallets function as vaults, keeping assets isolated and offline. Hardware wallets act as guards, requiring physical confirmation before funds can move. Transaction monitoring and verification serve as surveillance, catching errors before they become irreversible losses.

The uncomfortable truth is this: If you do not protect your crypto, no one else will.

There is no fraud department. No chargeback. No recovery desk. In a self-custody system, responsibility is absolute. The upside is sovereignty—the downside is that mistakes are final.

A layered security mindset doesn't make theft impossible. It makes it impractical. Hackers and scammers are opportunists. If you are difficult, slow, and cautious, they will move on.

Stay cautious. Stay educated. Stay secure.

BEST SECURITY PRACTICES: MAXIMUM PROTECTION FOR YOUR CRYPTO ASSETS

At first glance, comprehensive crypto security can feel overwhelming. In practice, once these systems are set up, they add very little friction to everyday use. The methods outlined below have been used successfully for years by experienced holders and traders. Strong security does not have to mean inconvenience—it means intentional design.

This level of protection is strongly recommended for anyone holding $10,000 or more in cryptocurrency. The goal is not paranoia. The goal is to remove single points of failure and make unauthorized access as difficult as possible.

COMPUTER & NETWORK SECURITY

Your devices and network form the foundation of your security posture. If they are compromised, everything above them is at risk.

A strong baseline includes:

- Using a dedicated device for crypto activity, separate from daily browsing, email, and research
- Installing reputable anti-virus and anti-malware software on all crypto-related devices
- Using a password vault to generate and store complex, unique passwords
- Separating browsers—one exclusively for crypto, another for general use
- Routing traffic through a VPN to reduce tracking and exposure
- Enabling biometric authentication and strong device passwords
- Securing Wi-Fi with WPA2 or WPA3 encryption and a strong passphrase
- Protecting the home network with a hardware firewall

- Using encrypted email for all crypto-related accounts

This layer reduces exposure to malware, phishing, traffic monitoring, and credential theft—some of the most common attack vectors in crypto.

WEBSITE & EXCHANGE SECURITY

Many losses occur not because wallets fail, but because users are tricked into trusting the wrong site or process.

Best practices include:

- Enabling two-factor authentication for every login and withdrawal
- Using hardware security keys where supported instead of SMS-based 2FA
- Requesting SIM-swap protection from your mobile provider
- Bookmarking official exchange and DeFi sites and accessing them directly
- Disconnecting and locking wallets after every session

The most dangerous moment in crypto is often not a hack—it is a rushed decision on a fake website.

WALLET SECURITY

Wallet security is the heart of self-custody. Everything ultimately comes down to how well private keys and seed phrases are protected.

Strong habits include:

- Never sharing seed phrases or private keys—under any circumstances
- Using multiple wallets instead of consolidating all assets in one place
- Storing multiple seed backups in physically separate, secure locations

- Using fireproof storage and durable metal backups for recovery phrases
- Locking wallets immediately after transactions
- Combining software wallets with hardware wallets for signing

This approach ensures that no single failure—device loss, malware, or phishing—results in total loss.

TRANSACTION & TRADING SECURITY

Even experienced users lose funds through simple, avoidable mistakes. Slowing down during transactions is one of the most effective protections available.

Safer transaction habits include:

- Copying and pasting wallet addresses rather than typing them
- Verifying the first and last characters of every address before sending
- Ignoring unknown tokens that appear unexpectedly in wallets
- Enabling separate 2FA for trading and withdrawals where available

Most irreversible losses happen in seconds. A few extra moments of verification can prevent them entirely.

FINAL THOUGHTS: CRYPTO SECURITY IS NON-NEGOTIABLE

No system can guarantee absolute safety—just as no car guarantees survival in every accident. But layered defenses dramatically reduce risk. Seatbelts, airbags, and brakes do not eliminate danger; they make survival far more likely.

Crypto security works the same way.

Casual investors benefit from strong passwords and 2FA.
Active traders benefit from hardware wallets and private networking.
Large holders require layered wallets, offline storage, and deliberate operational habits.

When these elements work together, they form something more than protection—they form resilience.

In a financial system where ownership is absolute and mistakes are permanent, security is not an accessory. It is the cost of sovereignty.

Build deliberately. Move carefully. And make yourself a target that attackers decide is not worth the effort.

CHAPTER 13 – YOU ARE YOUR BANK

FINAL THOUGHTS: SECURING YOUR FUTURE IN CRYPTOCURRENCY

Cryptocurrency represents one of the most significant financial shifts of our lifetime. It offers new ways to store value, transfer wealth, and participate in global markets without reliance on traditional institutions. For many, it opens the door to financial independence and long-term opportunity.

But cryptocurrency also removes the safety nets people have grown accustomed to.

- There is no bank fraud department.
- No account recovery hotline.
- No appeals process if funds are lost or stolen.
- In crypto, you are the bank.

That means you are also the custodian, the security officer, and the final authority over your assets. The freedom crypto provides is inseparable from the responsibility it demands. This book has outlined the tools, concepts, and practices needed to navigate that responsibility safely—but knowledge alone is not enough. Security only works when it is applied consistently and intentionally.

Self-custody is not about paranoia. It is about preparedness.

THE CORE LESSONS OF BEING YOUR OWN BANK

Throughout this book, several themes appear again and again—not because they are repetitive, but because they are foundational.

Centralized vs. Decentralized Systems

Centralized exchanges provide an accessible entry point. They are familiar, regulated, and convenient—but they require trust. When you use a centralized platform, you are temporarily giving up control of your private keys in exchange for ease of use.

Decentralized systems reverse that relationship. You retain full control, but with that control comes full responsibility. There is no intermediary to fix mistakes or reverse transactions.

For most people, the safest path is progressive:
start with centralized platforms, learn the mechanics, and move into decentralized systems only once strong security habits are in place.

WALLETS ARE THE CENTER OF SECURITY

In cryptocurrency, ownership is defined by control of private keys—not by accounts, usernames, or identity verification.

If you do not control the keys, you do not control the assets.

Using multiple wallets, separating long-term storage from daily use, and protecting recovery phrases offline are not advanced tactics—they are fundamental safeguards. Wallets are not accessories to crypto; they are the infrastructure of ownership.

SECURITY IS LAYERED, NOT SINGULAR

There is no single tool that makes crypto "safe."

Strong passwords alone are not enough.
Hardware wallets alone are not enough.
VPNs, antivirus software, and two-factor authentication each solve different problems.

Security works when layers overlap—when a failure in one area does not result in total loss. This defense-in-depth approach mirrors how banks, data centers, and critical infrastructure protect valuable assets. The difference is that you are now responsible for building it yourself.

SCAMS SUCCEED WHEN VIGILANCE FAILS

Most crypto losses are not the result of advanced hacks—they are the result of deception.

Phishing links, fake websites, impersonated support staff, rug pulls, and "too good to be true" opportunities all rely on one thing: urgency overriding caution.

Slowing down, verifying sources, double-checking transactions, and refusing unsolicited help eliminate the majority of real-world crypto threats. In a system where transactions are irreversible, patience is one of the most powerful security tools you have.

The Cost of Security vs. the Cost of Loss

Security always feels expensive—until it is compared to loss.

A few hundred dollars spent on hardware wallets, secure backups, and private networking may feel excessive at first. But compared to the permanent loss of tens of thousands of dollars—or more—it is a modest investment.

Many of the largest crypto losses in history were not caused by sophisticated attackers. They were caused by missing backups, exposed seed phrases, weak passwords, or blind trust. Nearly all of them were preventable.

Security is not about fear.
It is about foresight.

FINAL WORDS: OWNERSHIP IS THE POINT

Cryptocurrency is not just a new asset class—it is a new model of ownership. One that removes intermediaries, restores control, and places responsibility directly in the hands of the individual.

That responsibility cannot be outsourced.

Being your own bank means understanding your tools, respecting risk, and building systems that protect you even when you make mistakes. It means staying informed, questioning assumptions, and never relying on luck where preparation is possible.

This book has given you the framework. What comes next is practice.

- Build deliberately
- Move carefully
- Protect what you've earned

WELCOME TO THE ERA OF SELF-SOVEREIGN FINANCE—SECURE, INFORMED, AND FULLY IN YOUR CONTROL.

Made in the USA
Monee, IL
05 November 2026

67381959R00098